T0332223

Observational Molecular Astronomy

Exploring the Universe Using Molecular Line Emissions

Molecular line emissions offer researchers exciting opportunities to learn about the evolutionary state of the Milky Way and distant galaxies. This text provides a detailed introduction to molecular astrophysics and an array of useful techniques for observing astronomical phenomena at millimetre and submillimetre wavelengths. After discussing the theoretical underpinnings of molecular observation, the authors catalogue suitable molecular tracers for many types of astronomical regions in local and distant parts of the Universe, including cold gas reservoirs primed for the formation of new stars, regions of active star formation, giant photon-dominated regions, and near active galactic nuclei. Further chapters demonstrate how to obtain useful astronomical information from raw telescope data while providing recommendations for appropriate observing strategies. Replete with maps, charts, and references for further reading, this handbook will suit research astronomers and graduate students interested in broadening their skills to take advantage of the new facilities now coming online.

DAVID A. WILLIAMS is the Emeritus Perren Professor of Astronomy at University College London. A former president of the Royal Astronomical Society (2000–2002) and recipient of the RAS's Gold Medal (2009), he has led research groups in Manchester and London and has co-authored a number of texts on astrophysics and astrochemistry. His research interests centre on astrochemistry and using molecular line emissions to describe and understand the evolution of astronomical regions.

SERENA VITI is a professor of astrophysics at University College London. She began her career working on the spectroscopy of very cool stars but soon became interested in star formation and astrochemistry. She is the secretary of the European Astronomical Society and routinely serves on national and international scientific panels and committees.

Cambridge Observing Handbooks for Research Astronomers

Today's professional astronomers must be able to adapt to use telescopes and interpret data at all wavelengths. This series is designed to provide them with a collection of concise, self-contained handbooks, which cover the basic principles peculiar to observing in a particular spectral region, or to using a special technique or type of instrument. The books can be used as an introduction to the subject and as a handy reference for use at the telescope or in the office.

Series Editors
Professor Richard Ellis, *Department of Astronomy, California Institute of Technology*
Professor Steve Kahn, *Department of Physics, Stanford University*
Professor George Rieke, *Steward Observatory, University of Arizona, Tucson*
Dr. Peter B. Stetson, *Herzberg Institute of Astrophysics, Dominion Astrophysical Observatory, Victoria, British Columbia*

Books currently available in this series:

1. *Handbook of Infrared Astronomy*
   I. S. Glass

4. *Handbook of Pulsar Astronomy*
   D. R. Lorimer and M. Kramer

5. *Handbook of CCD Astronomy,* Second Edition
   Steve B. Howell

6. *Introduction to Astronomical Photometry*, Second Edition
   Edwin Budding and Osman Demircan

7. *Handbook of X-ray Astronomy*
   Edited by Keith Arnaud, Randall Smith, and Aneta Siemiginowska

8. *Practical Statistics for Astronomers*, Second Edition
   J. V. Wall and C. R. Jenkins

9. *Introduction to Astronomical Spectroscopy*
   Immo Appenzeller

10. *Observational Molecular Astronomy*
    David A. Williams and Serena Viti

Observational Molecular Astronomy

Exploring the Universe Using Molecular Line Emissions

DAVID A. WILLIAMS
University College London

SERENA VITI
University College London

CAMBRIDGE
UNIVERSITY PRESS

CAMBRIDGE
UNIVERSITY PRESS

Shaftesbury Road, Cambridge CB2 8EA, United Kingdom

One Liberty Plaza, 20th Floor, New York, NY 10006, USA

477 Williamstown Road, Port Melbourne, VIC 3207, Australia

314–321, 3rd Floor, Plot 3, Splendor Forum, Jasola District Centre, New Delhi – 110025, India

103 Penang Road, #05–06/07, Visioncrest Commercial, Singapore 238467

Cambridge University Press is part of Cambridge University Press & Assessment, a department of the University of Cambridge.

We share the University's mission to contribute to society through the pursuit of education, learning and research at the highest international levels of excellence.

www.cambridge.org
Information on this title: www.cambridge.org/9781107018167

First published 2013

A catalogue record for this publication is available from the British Library

Library of Congress Cataloging-in-Publication data
Williams, D. A. (David Arnold), 1937– author.
Observational molecular astronomy : exploring the universe using molecular line emissions / David A. Williams, University College London, Serena Viti, University College London.
pages   cm. – (Cambridge observing handbooks for research astronomers ; 10)
Includes bibliographical references and indexes.
ISBN 978-1-107-01816-7 (hard cover : alk. paper)
1. Molecular astrophysics.   I. Viti, Serena, author.   II. Title.   III. Series: Cambridge observing handbooks for research astronomers ; 10.
QB462.6.W55   2013
523'.02–dc23   2013013585

ISBN   978-1-107-01816-7   Hardback

Contents

      9.1   Chemical Modelling   152
      9.2   Radiative Transfer Modelling   158
      9.3   Further Reading   164

10   Observations: Which Molecule, Which Transition?   **165**
      10.1   Further Reading   167

      Appendix: Acronyms   169
      Index   171

List of Illustrations

List of Tables

List of Tables

Preface

This is a handbook for those astronomers who wish to use molecular line emissions as probes of astronomical sources. These sources may include molecular clouds and star-forming regions, circumstellar envelopes, and ejecta from evolved stars. Molecular lines are particularly useful in deconvolving complex emissions from distant unresolved galaxies.

This is not a textbook; it does not present detailed explanations and derivations. Textbook information can be found in the Further Reading sections at the end of each chapter. This handbook aims to provide a background of understanding so that the observer can begin to address the following questions:

- Why are different astronomical regions best traced in lines from different molecules?
- Which are the most suitable molecular tracers for studying the observer's selected sources?
- How does the observer convert raw telescope data into astrophysically useful information?
- How can the most complete physical description be extracted from the data?

1

Introduction

1.1 Why Are Molecules Important in Astronomy?

Molecules pervade the cooler, denser parts of the Universe. As a useful rule of thumb, cosmic gases at temperatures of less than a few thousand K and with number densities greater than one hydrogen atom per cm^3 are likely to contain some molecules; even the Sun's atmosphere is very slightly molecular in sunspots (where the temperature – at about 3200 K – is lower than the average surface temperature). However, if the gas kinetic temperatures are much lower, say about 100 K or less, and gas number densities much higher, say more than about 1000 hydrogen atoms per cm^3, the gas will usually be almost entirely molecular. The Giant Molecular Clouds (GMCs) in the Milky Way and in other spiral galaxies are clear examples of regions that are almost entirely molecular. The denser, cooler components of cosmic gas, such as the GMCs in the Milky Way Galaxy, contain a significant fraction of the nonstellar baryonic matter in the Galaxy. Counterparts of the GMCs in the Milky Way are found in nearby spiral galaxies (see Figure 1.1). Although molecular regions are generally relatively small in volume compared to hot gas in structures such as galactic jets or extended regions of very hot X-ray–emitting gas in interstellar space, their much higher density offsets that disparity, and so compact dense objects may be more massive than large tenuous regions.

Such dense, cool regions are of course important in themselves, in adding to our description of the total content of galaxies. But they are also important for our understanding of how galaxies evolve because this denser, cooler gas is the only reservoir of matter for future star formation. Measuring the mass of this reservoir gas in a galaxy and comparing with the existing stellar mass may, for example, give some indication of the evolutionary state of that galaxy. Alternatively, the interaction of an outflow from an active galactic nucleus with cool dense gas in the galaxy can produce a signature chemistry that through

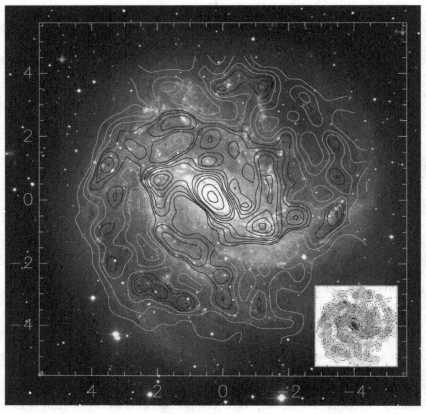

Figure 1.1. Velocity-integrated CO ($J = 1-0$) intensity as contours superposed on a map centre of M83 produced from images in B, V, and R of M83, a nearby spiral galaxy with an estimated mass of 10^{10} $M_\odot$. The x- and y-axes are offsets in RA and Dec from the centre of M83, measured in arcminutes. The average density in the GMCs is of the order of 100–400 cm^{-3}. The CO is associated with small regions of higher densities ($\leq 10^3$–10^5 cm^{-3}) and temperatures of the order of 50 K, where star formation occurs. The inset shows the CO (1–0) map in grayscale. (Reproduced with permission from Lundgren, A. A., Wiklind, T., Olofsson, H., and Rydbeck, G. 2004. *Astronomy & Astrophysics* 413, 505.) Copyright ESO.

its specific molecular emissions may reveal important details of the outflow, such as its mass loss rate. No less important, but on a smaller spatial scale than GMCs, the collapse of gas from a tenuous state to a dense star-forming core can be followed by measuring line emissions from the molecular gas, even though the temperature may be as low as 10 K or even less. Indeed, the low temperature is maintained during much of the collapse by these molecular emissions and also by continuum emissions of the dust. At the end of that collapse process, the newly formed star irradiates any surrounding debris that was not incorporated into the star and generates a new chemistry that provides new molecular signatures. In particular, a protoplanetary disk surrounding the young star is the location in which planet formation occurs, and is also almost entirely molecular. The disk responds to the intense and growing radiation from the central star, and to its powerful wind, in processes that can generate new and useful diagnostic molecules.

Thus, many processes of topical interest in modern astronomy and astrophysics involve cold dense gas or the interaction of radiation or of violent processes with cold dense gas. This book is offered as a guide for astronomers who wish to use molecules as probes of these kinds of processes, and in particular to address the following main questions:

- What kinds of molecules arise in different astronomical situations?
- Which molecular species are the most useful tracers of gas in these different situations?
- Which molecular species are the most useful for determining important physical parameters (e.g., cosmic ray flux, local radiation field, elemental abundances, and so forth) in those situations?
- How does one convert basic observational data taken at the telescope to astrophysically useful information (e.g., column densities or fractional abundances) about an astronomical object?

1.2 A Very Brief History of the Discovery of Molecules in Space

Optical absorption lines, apparently molecular in origin, were first detected in 1937 in the spectra of bright stars, along lines of sight through the diffuse interstellar medium of the Milky Way. A few years later, on the basis of laboratory work, these and other lines were attributed to the diatomic radicals CH, CH$^+$, and CN. No further detections were made until 1963, when OH masers were detected in the radio. Advances in detector technology permitted the development of millimetre wave and submillimetre wave astronomy and led to a veritable flood of new detections of molecular rotational transitions beginning in 1967. Some detections in other parts of the electromagnetic spectrum were also important. Molecular hydrogen, which has no dipole and therefore very weak rotational transitions, was first detected by a UV rocket experiment in 1970 by absorption in the Lyman and Werner bands; see Figure 1.2 for a

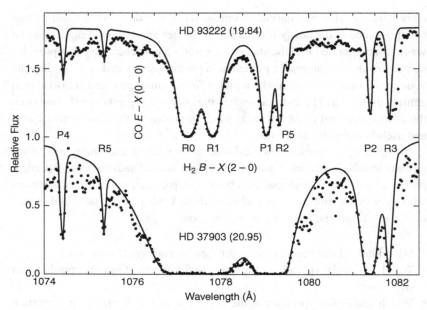

Figure 1.2. A piece of the UV absorption spectrum of H_2 towards two diffuse lines of sight taken with the space observatory Far Ultraviolet Spectroscopic Explorer (FUSE). These spectra show the electronic as well as ro-vibrational structure of the fundamental molecule H_2. The solid lines represent model fits to the spectra. The main features are the B–X (2–0) vibrational bands, and are labelled in the conventional notation (see Chapter 2). Here X represents the ground electronic state and B the first stable excited electronic state. The logarithm of the total hydrogen column density is indicated for each line of sight. (Reproduced by permission of the AAS from Sheffer, Y., Rogers, M., Federman, S. R., Abel, N. P., Gredel, R., Lambert, D. L., and Shaw, G. 2008. *Astrophysical Journal*, 687, 1075.)

recent detection. Molecular hydrogen is the seminal molecule for all interstellar chemistry, as we shall see.

From that time, the number of detected molecular species rose rapidly year by year and it soon became clear that the interstellar medium is a chemically complex environment (see e.g., Figure 1.3). An up-to-date list of detected molecular species is maintained at several websites (e.g., http://www.astro.uni-koeln.de/cdms/molecules/). A list of detected molecular species (as of 2012) organised by type of source is provided in Table 1.1.

Many isotopic varieties, in which, for example, D replaces H, or ^{13}C replaces ^{12}C, or ^{17}O or ^{18}O replaces ^{16}O, are also found, so that the total number of identified molecular species in interstellar and circumstellar space is very much larger than the total of main isotopes (which is currently ∼180).

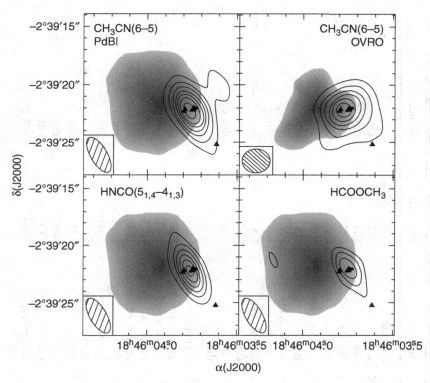

Figure 1.3. Spectra of complex molecules surrounding a massive star formation core, G29.96-0.02 (see also Chapter 5). The solid contours represent the molecular emissions and the grayscale indicates continuum emission from the ionised gas at 2.7 mm. (Reproduced, with permission, from Olmi, L., Cesaroni, R., Hofner, P., Kurtz, S., Churchwell, E., and Walmsley, C. M. 2003. *Astronomy & Astrophysics*, 407, 225.) Copyright ESO.

The first detections of extragalactic molecules were made in the 1970s. The current record for a molecular detection in a high-redshift galaxy is of CO at redshift $z = 6.42$ in 2003, in a gravitationally lensed quasar.

These detections attracted a great deal of attention, and the subjects of astrochemistry, bridging astronomy, chemistry, and physics emerged to try to account for the extraordinary range and variety of the detected species. However, there was no mere 'stamp-collecting phase' of molecular detections; in parallel with the development of astrochemistry, molecular emissions were immediately used to trace the existence of and physical conditions in interstellar and circumstellar gas. Such studies led to the discovery of previously unsuspected but important astronomical features – such as the GMCs in the inner part of the Milky Way. The structure of molecular outflows near cool stars

Table 1.1. List of detected molecular species with main regions in space where they have been observed

Molecule	Source	Molecule	Source	Molecule	Source
		2 Atoms			
H_2	dm, of	AlF	circ	AlCl	circ
CO	dm, circ, cc, yso, of, eg	C_2	dm	CH	dm, eg
SH	dm	CH^+	dm, eg	CN	dm, circ, eg
HCl^+	dm	CO^+	circ, yso	CP	circ
SiC	circ, yso	HCl	cc, yso, circ	KCl	circ
NH	dm, eg	NO	cc, eg	NS	dm, yso, eg
NaCl	circ	OH	dm, circ, eg	PN	yso
SO	cc, yso, dm, circ, of, eg	SO^+	dm, eg	SiN	circ
SiO	of, circ, yso, eg	SiS	of, circ, yso	CS	cc, yso, dm, circ, of, eg
HF	dm, eg	O_2	dm, yso	CF^+	dm
PO	circ	AlO	circ	OH^+	dm, eg
CN^-	circ	SH^+	dm		
		3 Atoms			
C_3	circ, dm	C_2H	yso, cc, dm, circ, eg	C_2O	cc
C_2S	cc, eg	CH_2	yso, dm	HCN	dm, cc, yso, circ, of, eg
HCO	cc, eg	HCO^+	cc, of, yso, eg	HCS^+	cc, yso, dm
HOC^+	eg	H_2O	cc, yso, of, circ, eg	H_2S	cc, yso, of, circ, eg
HNC	dm, cc, yso, circ, of, eg	HNO	yso	MgCN	circ
MgNC	circ	N_2H^+	cc, yso, of	N_2O	dm
NaCN	circ	OCS	cc, yso, of, eg	SO_2	cc, yso, of, eg
SiC_2	circ	CO_2	yso	NH_2	dm
H_3^+	dm	SiCN	circ	AlNC	circ
SiNC	circ	HCP	circ	CCP	circ
AlOH	circ	H_2O^+	dm, yso, eg	H_2Cl^+	dm, yso
KCN	circ	FeCN	circ		

6

4 Atoms

Molecule		Molecule		Molecule	
H_2O_2	dm	C_3H	cc, circ, eg	C_3N	dm
C_3O	dm	C_3S	ccc, circ	C_2H_2	cc, circ
NH_3	dm, cc, yso, of, eg	HCCN	circ	$HCNH^+$	cc
HNCO	yso, eg	HNCS	cc	$HOCO^+$	dm, yso
H_2CO	dm, cc, yso, of, eg	H_2CN	cc	H_2CS	cc, yso, of, circ, eg
H_3O^+	yso, eg	SiC_3	circ	CH_3	dm
C_3N^-	circ	HCNO	yso, eg	HOCN	dm, yso
HSCN	yso				

5 Atoms

Molecule		Molecule		Molecule	
C_5	circ	C_4H	dm, cc, circ	C_4Si	circ
C_3H_2	cc, yso, circ, eg	H_2CCN	cc, yso, circ	CH_4	circ
HC_3N	cc, circ, eg	HC(O)CN	yso	HCOOH	cc, yso
H_2CNH	dm, yso	H_2C_2O	dm, cc, yso	H_2NCN	yso
C_4H^-	circ	SiH_4	circ	H_2COH^+	dm, cc

6 Atoms

Molecule		Molecule		Molecule	
C_2H_4	circ	CH_3CN	cc, yso, of, eg	HC_4H	circ, eg
CH_3NC	yso	H_2C_4	circ, cc, yso	CH_2CNH	yso
C_5H	circ, cc	HC_3NH^+	cc	CH_3C_2H	cc, yso
C_5N	circ, cc	HC_4N	circ	CH_3OH	cc, yso, eg
HC_2CHO	yso	C_3H_2O		CH_3SH	yso
NH_2CHO	yso				

7 Atoms

Molecule		Molecule		Molecule	
C_6H	circ, cc, yso	C_6H	circ, cc, yso	CH_3NH_2	yso
C_2H_3CN	cc, yso, eg	HC_5N	circ, cc	CH_3CHO	cc, yso, eg
C_2H_3OH	yso	CH_2OCH_2	yso		

(cont.)

7

Table 1.1 (*cont.*)

Molecule	Source	Molecule	Source	Molecule	Source
			8 Atoms		
H_2C_6	circ, cc, yso	HC_6H	circ, eg	C_7H	circ, cc
CH_3C_3N	cc	CH_2CCHCN	cc	NH_2CH_2CN	yso
$HCOOCH_3$	yso, of	CH_3COOH	yso,	$HOCH_2CHO$	yso
C_2H_3CHO	yso				
			9 Atoms		
CH_3C_4H	cc	CH_3CHCH_2	cc	C_8H	circ, cc
HC_7N	circ, cc	C_8H	circ, cc	C_2H_5CN	yso
CH_3CONH_2	yso	C_2H_5OH	yso, of	CH_3OCH_3	yso
			10 Atoms		
CH_3C_5N	cc	CH_3COCH_3	yso	$HOCH_2CH_2OH$	yso
C_2H_5CHO	yso				
			11 Atoms		
CH_3C_6H	cc	HC_9N	circ, cc	$HCOOC_2H_5$	yso
			12 Atoms		
C_6H_6	circ, eg	C_3H_7CN	yso		
			13 Atoms		
$HC_{11}N$	circ, cc				

Abbreviations: dm = diffuse medium (including translucent clouds); circ = circumstellar envelope around evolved star/protoplanetary nebula; cc = cold cloud core; yso = gas around a young stellar object, including observations of the hot core in the galactic centre; of = outflow; eg = extragalactic regions. Some of the abbreviations used in this list are taken from E. Herbst and E. F. van Dishoeck. 2009. *Annual Review of Astronomy and Astrophysics, 47*: 427. We do not include isotopologues in this table.

8

Table 1.2. Types of interstellar and circumstellar region and their physical characteristics

Region	n_H (cm^{-3})	T (K)
Coronal gas	$<10^{-2}$	5×10^5
HII regions	>100	1×10^4
Diffuse gas	100–300	70
Molecular clouds	10^4	10
Prestellar cores	10^5–10^6	10–30
Star-forming regions	10^7–10^8	100–300
Protoplanetary disks	10^4(outer)–10^{10}(inner)	10(outer)–500(inner)
Envelopes of evolved stars	10^{10}	2000–3500

All of the regions, except coronal gas and HII regions, can be probed with molecules.

was revealed, and molecular ices were found to be present in the interstellar medium.

Molecular emissions, along with X-ray, UV, optical, and infrared emissions, have helped to define the variety of physical states of interstellar gas. These range over at least a factor of $\sim 10^{12}$ in density and $\sim 10^5$ in temperature, from number densities of $\sim 10^{-2}$ cm^{-3} and temperatures $\sim 10^6$ K in so-called coronal gas to values of $\sim 10^{10}$ cm^{-3} and ~ 10 K in protoplanetary disks. Table 1.2 lists the known interstellar and circumstellar components. Of these types of region, diffuse clouds, molecular clouds, prestellar cores, star-forming regions, protoplanetary disks, circumstellar envelopes, and the ejecta of novae and supernovae can be studied through molecular emissions.

As astronomy moves into a new phase dominated by data from revolutionary space- and ground-based instrumentation, molecular astronomy is no longer a semidetached specialty of work in the millimetre and submillimetre regions of the spectrum. Molecular astronomy now addresses questions at the forefront of the subject, and is simply part of the range of expertise that astronomers must command. This book is intended to help astronomers become equally skilled in molecular line observations as in making observations in other regions of the spectrum.

1.3 Gas and Dust

1.3.1 Gas Composition for Interstellar Chemistry

The raw material for our considerations of chemistry consists of gas and dust. The gas consists mainly of hydrogen and helium with a small component

Table 1.3. Approximate solar
elemental abundances relative to
the total number of hydrogen nuclei

Element	Abundance
H	1
He	9×10^{-2}
O	5×10^{-4}
C	3×10^{-4}
N	7×10^{-5}
Si	3×10^{-5}
Mg	4×10^{-5}
Fe	3×10^{-5}
S	1×10^{-5}
Na, Ca	2×10^{-6}

Note that solar elemental abundances may not be valid for all regions
of space.

of other elements formed in stellar nucleosynthesis and distributed by novae
and supernovae and by stellar winds. Obviously, the ability of a gas to form
molecules involving carbon, oxygen, nitrogen, sulfur, and other elements (as
well as hydrogen) depends on the abundance of the small component of other
elements relative to hydrogen. These relative elemental abundances may vary
from place to place within a galaxy and from galaxy to galaxy. Solar abundances
are often used as a conventional reference level; solar elemental abundances
relative to hydrogen are shown in Table 1.3. Gas with these relative elemental
abundances is said to have solar *metallicity*. The metallicity is an important
parameter in astrochemistry; we consider the effect on the chemistry of varying
the metallicity in Chapter 4. It is often assumed that although the metallicity
may vary, the abundances of the elements relative to each other follow solar values. However, this may not be the case everywhere. For example, if considering
the early Universe, supernovae of different masses may lead to quite different
predictions of relative abundances of the major elements carbon, nitrogen, and
oxygen. Stellar evolution models for initially zero-metallicity gas predict nitrogen to be underabundant whereas oxygen and magnesium are overabundant
compared to solar metallicity. Also, dredge-up processes in evolved stars are
observed to create distinct differences in elemental abundances. For example,
some stars may have different C:O ratios in their atmospheres and envelopes

at different evolutionary stages. There are examples of a stellar atmosphere being at one stage carbon-rich and at another oxygen-rich. Thus, the relative elemental abundances to be used are not always solar.

1.3.2 Dust

Dust is observed to be mixed with interstellar gas in all galaxies. It is detected either through the extinction that it causes at UV, optical, and infrared wavelengths (see Figure 1.4), or by the detection of thermal emission from warm dust in the vicinity of stars and from cooler dust in dark clouds. Dust is also present in the envelopes of cool stars and of novae and supernovae; these are the locations where dust is believed to nucleate and grow. The observations on long low-density paths in the Milky Way Galaxy require that dust grains range in size, a, from about a nanometre to about a micron, with number density $n_d(a)da$ in the range a to $a + da$ heavily weighted to the smaller grains; $n_d(a)da \sim a^{-3.5}da$. The grains are asymmetric in shape, and the larger grains (at least) can be partially aligned by the local magnetic field. The smallest grains may be molecular (rather than bulk) in nature and may include polycyclic aromatic hydrocarbons (PAHs) with some graphitic-type structure. The chemical composition of grains includes carbons of both ring and polymeric structures, amorphous and crystalline silicates, and probably various other metallic oxides.

From the perspective of this book, we shall be concerned with the various roles of dust in the interstellar medium. Its primary role is to extinguish UV and visual radiation in the interior of gas clouds, thereby shielding interior material from the destructive effects of starlight. Some of the absorbed radiation releases photoelectrons from the dust grains; these are an important energy source for the gas. Optical and UV starlight that is absorbed heats the grains, which then radiate in the far-IR (see Figure 1.5). Another important role of dust is in catalysing reactions on grain surfaces, especially the formation of molecular hydrogen, and contributing product molecules to the network of gas reactions. In some darker regions, icy mantles accumulate on the surfaces of dust grains; these mantles are sinks for molecules, removing them from the gas. The mantles are observed to contain water, carbon dioxide, carbon monoxide, methanol, and other species. Chemical processing can also be formed in these ices, making molecular species of greater complexity than can easily occur in the gas phase. Thermal and nonthermal processes may return the ices to the gas phase. Dust grains also tend to 'mop up' electrons from the gas, thereby affecting the gaseous ionisation level (and consequently the gas-phase chemistry) within interstellar clouds.

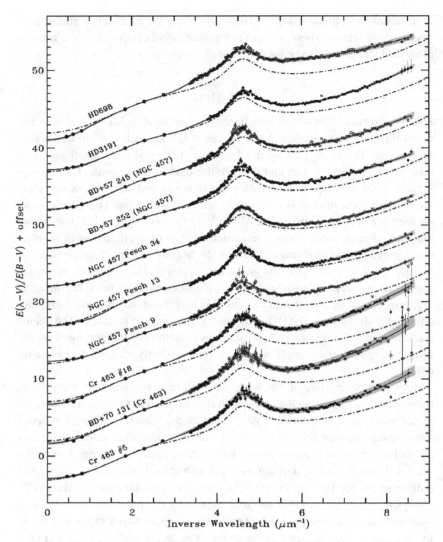

Figure 1.4. Interstellar extinction curves (offset for clarity) for lines of sight towards a number of bright stars in the Milky Way. The curves are normalised, but they all have the same basic shape – higher extinction in the UV than in the IR and a near linear part in the visual. This requires a distribution of grain sizes, with many more small grains than large. The dot-dash curve is the mean interstellar extinction curve. (Reproduced by permission of the AAS from Fitzpatrick, E. L., and Massa, D. 2007. *Astrophysical Journal*, 663, 320.)

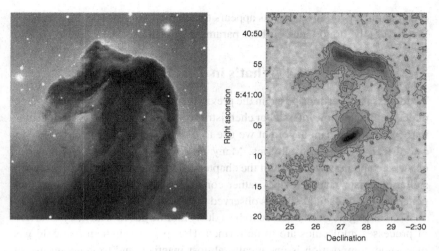

Figure 1.5. The Horsehead Nebula (left) optical image from the VLT (ESO) and (right) the JCMT-SCUBA 850 μm contour map. The bright submillimetre regions indicate emission from prestellar objects. (Reproduced with permission from Ward-Thompson, D., Notter, D., Bontemps, S., Whitworth, A., and Attwood, R. 2006. *Monthly Notices of the Royal Astronomical Society*, 369, 1201.)

Although the chemical nature and physical structure of the dust are undoubtedly important for all these topics, we do not discuss them here. We do not describe other dust properties, nor the origin of dust in supernovae, novae, and other stellar winds, but simply refer to relevant works on the subject (see Further Reading).

However, it is important to note that the grains lock up a significant fraction of certain elements; for example, models of interstellar extinction in the Milky Way Galaxy require that almost all of the element silicon is in silicate dust, whereas a significant fraction of carbon, up to perhaps one half, is in carbon dust. Clearly, those atoms locked in the dust are not available for gas-phase chemistry unless the grains are being eroded in high-temperature gas. If not, then the solar abundances shown in Table 1.3 are not wholly available for chemistry in the gas phase and need to be adjusted to allow for the components locked in the dust.

Finally, like metallicity, the dust:gas ratio is not a fixed parameter. The quantity of dust is a consequence of the history of stellar evolution, so may vary within and between galaxies. Although the ratio is difficult to determine accurately, it is probably related to the metallicity and may vary from one galaxy to another, or even within one galaxy. The dust:gas ratio in nearby external galaxies appears to correlate fairly well with metallicity. The range in each parameter for these objects appears to be around one order of magnitude. We discuss the importance of these parameters in Chapter 4.

1.4 What's in This Book

To address the questions posed in the previous sections, astronomers need some understanding of how interstellar chemistry works. Chapter 2 gives a brief summary of the basic language that we use to discuss molecular spectroscopy as a tool in molecular astrophysics. Many readers will have met this material elsewhere and may wish to omit the chapter. Then Chapter 3 summarizes interstellar chemistry and describes rather concisely the network of reactions that generates the molecular species observed (and many that are not yet detected).

The starting point is assumed to be a dusty atomic gas in which, by a variety of processes, molecules are to be formed. However, an undisturbed cold gas of cosmic composition is chemically almost inactive, and to generate useful tracer molecules on a reasonable timescale an efficient chemistry needs to be switched into action. The necessary switch, or *driver*, may be starlight, cosmic rays, surface reactions on dust grains, or gas dynamical processes. Both starlight and cosmic rays can act as sources of energy capable of generating ionisation and heat in the gas so that a chemistry may be initiated. Grain surfaces can catalyse new products from atoms and molecules arriving at the surface. Gas dynamical processes such as shocks or turbulent interfaces may introduce heat and possibly ionisation into the gas, initiating a characteristic chemistry.

In Chapter 4 the sensitivity of the chemistry to particular influences is explored. Do characteristic 'signature' molecules arise when a particular driver of the chemistry is enhanced above a conventional level? We look in particular at regions in which the stellar radiation is especially powerful, at dense cold regions, and at turbulent interfaces between cold and hot moving gases. We also ask: What happens to the chemistry if the dust:gas ratio or the elemental abundances depart significantly from canonical values?

Chapter 5 applies the ideas developed in the previous chapters to several important and much-studied situations in the Milky Way: these are molecular clouds and star-forming regions where either low-mass or high-mass stars are being formed. For each of these regions, the main question posed is this: What are the most appropriate tracer molecules to use in probing these situations?

In Chapter 6 we apply the ideas developed so far to interstellar gas in external galaxies. In these situations, a much wider range of physical conditions may occur compared to those in the Milky Way because the drivers of the chemistry, such as cosmic ray flux or starlight intensity, the dust:gas ratio, or the elemental abundances, may themselves be very different from Milky Way values. What are the molecules that allow observers to probe those drivers? Can we infer values of those physical quantities in distant galaxies?

In Chapter 7 we peer into the early Universe and review the role of molecules in pregalactic astronomy, and we speculate on the possibility of detecting molecules during the very early events of the history of the Universe.

It is, however, a significant step from raw data taken at the telescope to information on, say, molecular column densities that is astronomically useful. Chapter 8 describes the conventional approaches that allow the observer to convert data from the telescope to useful measures such as molecular column densities or fractional abundances and to obtain measures of density and temperature. Chapter 9 summarises current numerical approaches to radiative transfer. From measures such as these the astronomer may infer masses and begin to make useful speculations. Finally, Chapter 10 provides information for some molecular transitions often used as tracers of different types of interstellar and circumstellar regions.

1.5 Further Reading

Cernicharo, J., and Bachiller, R., eds. 2011. *The Molecular Universe*. IAU Symposium 280. Cambridge: Cambridge University Press.

Draine, Bruce T. 2011. *Physics of the Interstellar and Intergalactic Medium*. Princeton Series in Astrophysics. Princeton, NJ: Princeton University Press.

Hartquist, T. W., and Williams, D. A., eds. 1998. *The Molecular Astrophysics of Stars and Galaxies*. New York: Oxford University Press.

Tielens, A. G. G. M. 2005. *The Physics and Chemistry of the Interstellar Medium*. Cambridge: Cambridge University Press.

Whittet, D. C. B. 2003. *Dust in the Galactic Environment*. Bristol UK: Institute of Physics Publishing.

2

Spectra and Excitation of Interstellar Molecules

In this chapter we summarise briefly the notation of molecular spectroscopy, with examples of transitions used to identify molecular species in the interstellar medium. We also describe how radiation is transported in the interstellar medium, introducing ideas that will be needed in Chapters 8 and 9. Finally, we discuss the processes that determine level populations of molecules in the interstellar medium.

2.1 Molecular Spectroscopy

Whereas most atomic spectra are determined simply by transitions between individual electronic states, molecular spectra are more complex because molecules have additional degrees of freedom associated with vibration and rotation. Each electronic state of a molecule possesses a manifold of vibrational levels, and each of those vibrational levels has a ladder of rotational levels associated with it. Electronic transitions in molecules therefore occur between specific vibrational and specific rotational levels in each electronic state. The equivalent of a single atomic line corresponding to an electronic transition is – for a molecule – replaced by a set of many lines (see Figure 1.2 for part of the H_2 electronic transition B–X, between the two lowest electronic states, showing the vibrational and rotational structure). These electronic transitions often lie in the ultraviolet spectrum.

Molecules also possess transitions that have no counterpart in atoms. Molecules may undergo transitions between specific rotational states of vibrational states, that is, ro-vibrational transitions. These transitions usually occur in the near-infrared range. Pure rotational transitions may also occur between rotational states of the same vibrational level. These tend to lie in the far-infrared, or in the millimetre to submillimetre range of the electromagnetic spectrum.

16

Molecules in the interstellar medium may be identified by electronic, rovibrational, or pure rotational transitions. By far the most detections have been made in the millimetre and submillimetre range, corresponding to pure rotational transitions, almost always of molecules in their ground electronic state and lowest vibrational level. There are also some other types of transition (e.g., in OH and NH_3) that do not fit this pattern. In general, if the potential energy curve of a molecule has a minimum, then the energies of the discrete rovibrational levels within an electronic state can be determined by the use of the Born–Oppenheimer approximation, which allows us to treat the electronic and nuclear motions separately. In most astronomical contexts this approximation is valid.

In this chapter we describe briefly the spectroscopic notation for diatomic and polyatomic molecules and summarise the important mechanisms that affect the excitation of molecules.

2.1.1 Diatomic Molecules

Electronic states are labelled according to their electronic spin and the magnitude of the component of the electronic angular momentum along the internuclear axis, Λ. The first three electronic states are Σ, Π, and Δ. Homonuclear diatomics are either symmetric (g) or antisymmetric (u) with respect to the interchange of the two nuclei; $-$ and $+$ superscripts are used to indicate that the electronic state does or does not change sign when reflected at the origin, respectively. So, for example, the ground electronic state of H_2 has a spin of 0 and is labeled $^1\Sigma_g^+$.

The nuclear motion can be separated into its vibrational and rotational components. The discrete vibrational levels are denoted by $v = 0, 1, 2, \ldots$ usually with an energy difference of about 0.2 eV (although note that for H_2 this difference is about 0.5 eV). The rotational components are labelled $N = 0, 1, 2 \ldots$ with an energy separation of $2BN$, where N is the rotational quantum number of the higher rotational level and B is the rotational constant, equal to $\hbar^2/2I$, where I is the moment of inertia. For carbon monoxide, CO, $2B$ is equivalent to $\sim 5\,\text{K}$.

In hydrogen-bearing molecules, when N is even, the sum of the nuclear spins must be 0 and the state is said to be *para-* whereas when N is odd the sum of the nuclear spins must be 1 and the state is *ortho-*. Transitions between *ortho-* and *para-* states are forbidden.

In the cold interstellar medium ($T < 100\,\text{K}$), most of the observed rotational transitions belong to the ground electronic states of molecules and lowest vibrational level. For some common interstellar molecules (e.g., CO) this is the $^1\Sigma$ state and in this case the sum, J, of the rotational angular momentum, the electronic angular momentum, and the electronic spin is just the rotational angular momentum and therefore $J = N$ and defines the rotational level of the molecule. However, for other important diatomics such as OH the ground electronic state can be a $^2\Pi$; in this case interactions between spins and different angular momenta are important. For example, the interaction between J and $\mathcal{I}$, the total spin of the nuclei, causes hyperfine splitting. A common example in astronomy is again OH at 18 cm, where the strongest emission indeed comes from the $F = 1 \to 2$ hyperfine component of OH, where F is the quantum number associated with the hyperfine splitting.

For a transition to occur, the electric dipole moment, μ, must change when the atoms are displaced relative to one another. This happens only for heteronuclear diatomic molecules because homonuclear molecules have the charges distributed symmetrically: vibrational and rotational spectra of homonuclear diatomics can be obtained only if an electronic transition also occurs.

Selection rules for electronic transitions in heteronuclear diatomic molecules are as follows: $\Delta\Lambda = 0, \pm 1$, $g \longleftrightarrow u$; the rotational selection rules are $\Delta J = J(\text{upper})-J(\text{lower}) = \pm 1$ for diatomics with no net spin or orbital angular momentum; otherwise $\Delta J = 0$ is also possible (but not for the $\Sigma \longleftrightarrow \Sigma$ electronic transitions for which $\Delta J = 0$ rotational transitions are forbidden). Homonuclear diatomics selection rules include $\Delta J = \pm 2$. $\Delta J = -2, -1, 0, 1, 2$ are labelled O, P, Q, R, and S branches.

Examples of diatomic transitions relevant to interstellar studies are H_2 ($J = 2 \to 0$ and $3 \to 1$) at wavelengths of 28 and 16 μm, respectively, and the CO ($J = 1 \to 0$, $J = 2 \to 1$) at frequencies of 115.271 and 230.538 GHz, respectively.

2.1.2 Polyatomic Molecules

Polyatomic molecules can be classified into three different categories: linear molecules, symmetric tops, and asymmetric tops. We discuss them separately for simplicity.

Linear Molecules

For many linear molecules the total angular momentum minus the nuclear spin J is equal to N; in this case the energy levels of the lowest vibrational level of the ground electronic state are given by $\sim BJ(J + 1)$ where B is, as before, a rotational constant equal to $\hbar^2/2I$, and J is used to specify the rotational level of the molecule. Because B decreases as the mass increases, the energy levels tend to be closer (and hence the wavelength is larger) the heavier the molecule. In practice this means that in the submillimetre spectrum, for heavy species, several transitions are found close together.

Vibrations can occur in a 'stretching mode', which is along the length of the molecule; and in a 'bending mode', which is perpendicular to it, and is specified by two quantum numbers, one for the radial coordinate and one for the angular coordinate.

Selection rules for linear molecules are $\Delta J = \pm 1$, with, as before, $\Delta J = -1$ transitions giving rise to the lower energy branch (P-branch), and the $\Delta J = +1$ transitions giving rise to the higher energy branch (R-branch).

Examples of transitions of linear molecules relevant to interstellar studies are HCN ($J = 3 \rightarrow 2$) at 265.886 GHz and HCO^+ ($J = 1 \rightarrow 0$) at 89.1885 GHz.

Symmetric Tops

Symmetric tops are those nonlinear polyatomic molecules for which two of the moments of inertia, I_A, I_B, I_C (where A, B, C refer to mutually orthogonal axes), are the same. We can divide symmetric tops into two categories: oblate and prolate, with (respectively) $I_A = I_B < I_C$, and $I_A < I_B = I_C$. Note that spherical top molecules such as CH_4 have equal moments of inertia but no dipole moments, which means that transitions between rotational levels in the ground electronic states are weak. The energies of a symmetric top are given by:

$$E(J, K) = \frac{\hbar^2}{2I_A} J(J + 1) + \frac{\hbar^2}{2} \left\{ \frac{1}{I_A} - \frac{1}{I_B} \right\} K^2 \qquad (2.1)$$

where K is the projection of J onto the principal axis. The inversion properties of symmetric top molecules are often important for astrophysical spectroscopy; for example, in the case of ammonia, considerations of symmetry and nuclear statistics show that $K = 0$ levels are not split into inversion sub-levels whereas all others (J, K) are. Observed NH_3 inversion transitions are at wavelengths around 1.3 cm.

The selection rules for symmetric top molecules are $\Delta K = 0$ and $\Delta J = 0, \pm 1$.

Examples of transitions of symmetric tops relevant to interstellar studies are the H_3^+ $J = 0 \rightarrow 1$ transition at a wavelength of 3.66 μm and the multi-transitions of the $J = 12 \rightarrow 11$ of the CH_3CN molecules at frequencies in the 220.59–220.74 GHz range.

Asymmetric Tops

Asymmetric tops are nonlinear rotors; their moments of inertia are all different and the energy levels have to be computed numerically for a suitable potential energy surface or determined in the laboratory. The energy levels are specified with the quantum numbers J_{K_A, K_C}, where K_A and K_C are the projections of J on the A and C principal axes, respectively.

The selection rules for asymmetric tops depend on the components of the permanent dipole moment, μ, along the A, B, and C principal molecular axes. A simplified general scheme is as follows: when $\mu_A \neq 0$, $\Delta K_A = 0, \pm 2, \pm 4 \ldots$, $\Delta K_C = \pm 1, \pm 3, \ldots$; when $\mu_B \neq 0$, $\Delta K_A = \pm 1, \pm 3 \ldots$, $\Delta K_C = \pm 1, \pm 3, \ldots$; when $\mu_C \neq 0$, $\Delta K_A = \pm 1, \pm 3 \ldots$, $\Delta K_C = 0, \pm 2, \pm 4, \ldots$

Examples of asymmetric top transitions relevant to interstellar studies are the fundamental *ortho*-H_2O (1_{10}–1_{01}) transition at a wavelength of 557 GHz, the *para*-H_2CO (2_{02}–1_{01}) transition at a frequency of 145.60 GHz, and the HCOOH ($10_{3,7}$–$9_{3,6}$) transition at 225.51 GHz.

There are several transitions of molecules that are important in the interstellar medium whose spectroscopy is not covered by the simple description and rules listed in the preceding text. We here list two examples of some relevance to interstellar studies.

NH_3 Inversion: Quantum mechanical tunnelling allows the nitrogen atom at the apex of the pyramidal shape of the molecule of ammonia to pass through the pyramid base to the other side, reversing the orientation of the pyramid: the transition that comes from this motion is 'inverted' with respect to a standard rotation. Several inverted transitions in the centimetre range are observed in the interstellar medium, with the most common being the $(J, K) = (1, 1)$ and $(2, 2)$ lines at 23.694 and 23.722 GHz, respectively.

OH radio lines: If the width of an absorption or emission line is larger than the hyperfine splittings of that particular state then the hyperfine levels will be 'sharing' the same photons and hence the hyperfine levels will interact. This process is called radiative relaxation between two hyperfine levels. For OH this has been observed in radio at several frequencies including the Λ-doublet transitions of the $^2\Pi_{3/2}(J = 3/2)$ at 18 cm, the $^2\Pi_{1/2}(J = 1/2)$ at 6.3 cm, etc.

2.2 Radiative Transport in the Interstellar Medium

In Chapter 8 we look more closely at methods of determining the intensity of a transition from observations but here it is worth recalling how the specific intensity of radiation can be estimated as a function of position and frequency;

this is done using the Radiative Transfer Equation:

$$\frac{dI_\nu}{d\tau_\nu} = \frac{j_\nu}{\alpha_\nu} - I_\nu = S_\nu - I_\nu \tag{2.2}$$

and its general solution:

$$I_\nu(\tau_\nu) = I_\nu(0)e^{-\tau_\nu} + \int_0^{\tau_\nu} e^{-(\tau_\nu - \tau_\nu')} S_\nu(\tau_\nu') \, d\tau_\nu' \tag{2.3}$$

where $I_\nu(\tau_\nu)$ is the specific intensity as a function of the path; τ_ν is the optical depth; and S_ν is the source function, which is the ratio between the emission (j_ν) and absorption (α_ν) coefficients. In the general solution, the first term describes the absorption of the incident radiation $I_\nu(0)$, while in the second term we have an integration over the emitted photons from the source function, as well as a factor to include absorption of the emitted photons propagating to optical depth τ_ν.

It is worth describing the solutions for the two limiting cases, which apply if the source function is constant in the region of interest:

(1) the optically thin case, where τ_ν approaches 0; then $I_\nu(\tau_\nu) \simeq I_\nu(0) + S_\nu(\tau_\nu)$;
(2) the optically thick case, where τ_ν approaches ∞ and $I_\nu(\tau_\nu)$ approaches S_ν.

An example of an optically thin region in the interstellar medium may be that of a hot, low-density nebula (e.g., NGC 7009), and an example of an optically thick region in the interstellar medium may be clumpy molecular clouds (e.g., L1673).

2.3 Determining the Level Populations

As we see in Chapter 3, molecular transitions can be induced by several processes. These may involve interactions with photons, cosmic rays, or collisions with other molecules, atoms, or free electrons. In this section we briefly summarise what happens to the atoms or molecules during such processes:

Excitation and de-excitation: Excitation can be radiative or collisional; in both cases an electron jumps from a lower energy level to a higher one, where the extra energy is provided by the colliding atom/molecule or by a photon that has an energy corresponding to the energy difference between the two energy levels, respectively. In a molecular cloud the collisions usually occur with molecular hydrogen or with helium. In the case of a radiative excitation, once an atom or molecule is excited to a higher level, it can jump back to the lower level, emitting radiation from the upper level; the two states are bound states in which the electron is bound to the atom and the emitted photon is at a particular energy and wavelength (line emission). The emission is spontaneous and occurs at a rate that is determined by the Einstein A-coefficient (radiative de-excitation), A_{ji} (s^{-1}), or transition probability for spontaneous decay from an upper state j to a lower state i, with the emission of a photon (radiative decay); the time taken for an electron in state j to spontaneously decay to state i is, on average, $1/A_{ji}$. If during the transition of the electron from a lower energy level to a higher one a photon is absorbed then an *absorption* line is formed. The rate is then determined by B_{ji} (s^{-1} J^{-1} m^2 sr), the Einstein coefficient for radiatively induced de-excitation from an upper state to a lower state.

However, collisional de-excitation can also occur; if so, no radiation is emitted.

Above a certain critical density, which varies from transition to transition, the population is thermalised; in other words, one can define the critical density as the density at which the rate of upward transitions through collisions is equal to the rate of downward transitions through spontaneous decay. The critical density, n_{crit}, for a specific transition is therefore equal to A_{ji}/γ_{ji} where γ_{ji} is the rate coefficient for collisional de-excitation from j to i. At densities higher than the critical density, the upper state is normally de-excited in collisions. At densities lower than the critical density, the population in the upper state is not highly populated. But at densities near the critical density, maximum emission occurs. Hence, each transition will tend to trace gas at a particular density near the critical density for that transition, and one can use different molecules and transitions to map different parts of a cloud where density is not uniform, as we see in later chapters. An example of excitation followed by de-excitation is CO in its ground state colliding with H_2:

$$CO(J = 0, 1) + H_2 \rightarrow CO(J = 2) + H_2 \tag{2.4}$$

$$CO(J = 2) \rightarrow CO(J = 0, 1) + h\nu \tag{2.5}$$

Ionisation and recombination: Ionisation of an atom or a molecule may be caused by absorption of radiation or by collision with a cosmic ray or other fast particle. In the case of radiation, the absorption spectrum is a continuum. An example of photoionisation is:

$$CH + h\nu \rightarrow CH^+ + e \tag{2.6}$$

where a photon, $h\nu$, from an incident radiation field is absorbed and causes the ionisation. In this example, absorption occurs for photons with energies above the ionisation potential of CH, which is 3.47 eV.

The reverse process, in which a positive ion captures an electron, is called recombination. For example, in the recombination:

$$H_3O^+ + e \rightarrow H_2O + H \tag{2.7}$$

the energy released when the ion and electron recombine is so great that the complex dissociates.

Molecular lines can appear in emission or in absorption, depending on the excitation temperature, T_{ex}, of the transition, which is the temperature that gives the observed ratio of two energy levels in a Boltzmann distribution (see Chapter 8). If T_{ex} differs from the brightness temperature, T_B, of the radiation at the frequency of the transition then the line appears, either in emission ($T_{ex} > T_B$) or in absorption ($T_{ex} < T_B$). The excitation temperature is dependent on the critical density of a particular transition; different molecular lines have different critical densities, n_{crit}. So, for regions with density near the critical density for a particular transition, the line becomes 'visible' and may be used as an approximate density diagnostic. In these cases T_{ex} is close enough to the kinetic temperature of the gas to be used as a temperature indicator.

2.4 Further Reading

Draine, B. T. *Physics of the Interstellar and Intergalactic Medium*. Princeton Series in Astrophysics. Princeton, NJ: Princeton University Press.

Hartquist, T. W., and Williams, D. A. 1998. *The Molecular Astrophysics of Stars and Galaxies*. eds. New York: Oxford University Press.

Tennyson, J. 2005. *Astronomical Spectroscopy: An Introduction to the Atomic and Molecular Physics of Astronomical Spectra*. London: Imperial College Press.

3

Astrochemical Processes

3.1 What Drives Cosmic Chemistry?

It is a relatively straightforward matter to use freely available computer codes and lists of chemical reactions to compute abundances of molecular species for many types of interstellar or circumstellar region. For example, the UDfA, Ohio, and KIDA websites (see Chapter 9) provide lists of relevant chemical reactions and reaction rate data. Codes to integrate time-dependent chemical rate equations incorporating these data are widely available and provide as outputs the chemical abundances as functions of time. For many circumstances, the codes are fast, and the reaction rate data (from laboratory experiments and from theory) have been assessed for accuracy. The required input data define the relevant physical conditions for the region to be investigated.

These codes and databases are immensely useful achievements that are based on decades of research. However, the results from this approach do not readily provide the insight that addresses some of the questions we posed in Chapter 1: What are the useful molecular tracers for observers to use, and how do these tracers respond to changes in the 'drivers' of the chemistry? Observers do not need to understand all the details of the chemical networks (which may contain thousands of reactions), but it is important to appreciate how the choice of the tracer molecule may be guided by, and depend on, the physical conditions in the regions they wish to study.

We use the word 'driver' to indicate that, as discussed in Section 1.4, interstellar chemistry needs to 'driven'. This may be done by creating rapid reaction paths, by catalyzing reactions between reactants, or by providing a source of energy that heats the gas. In general, a gas at low density and low temperature, of cosmic composition, left to its own devices, will not generate an extensive chemistry that produces useful tracer molecules. The chemistry must be 'driven', and the 'drivers' available in the interstellar medium are starlight, cosmic rays, grain catalysis, and gas dynamics. Each driver produces a characteristic chemistry from which useful molecular tracers may be selected. The first three drivers create new and efficient chemical pathways, while the third injects energy into the gas to allow barriers to reaction to be overcome. For example, diffuse clouds, such as those for which H_2 spectra are shown in Figure 1.2, are readily penetrated by starlight; this is the main driver of chemistry in such regions. Starlight is largely excluded from molecular clouds, such as Barnard 68, illustrated in Figure 3.1 top, but cosmic rays are almost unimpeded as they pass through such regions, and they drive much of the chemistry. Grain surfaces become important drivers of complex chemistry in very dense regions of massive star formation, illustrated in Figure 1.3. Finally, gas dynamics in regions such as the stellar jet HH49/50, illustrated in Figure 3.1 bottom, injects heat and drives a characteristic chemistry in the interface between the jet and the ambient gas.

Of course, in many situations some or all of the drivers may be acting together, while in others the chemistry may be dominated by one driver, such as a very intense UV field in a region of massive star formation or an unusually high flux of cosmic rays generated in energetic regions of an active galaxy. For clarity, we first discuss the chemistry arising from each driver separately, even though this dependence on a single driver is likely to be unrealistic. For present purposes, we assume that the gas in which the chemistry is to be promoted is composed of hydrogen atoms and molecules, helium, and other elements as neutral atoms in proportion to the solar abundances (see Table 1.3).

3.2 Chemistry Initiated by Electromagnetic Radiation

3.2.1 Photon-Dominated Regions

The interstellar radiation field impinging on an interstellar cloud is generally dominated by the contribution from O and B stars. Radiation in the UV is strongly attenuated by dust absorption, and at a depth into a cloud equivalent to A_V magnitudes the intensity is reduced by a factor of about e^{2A_V}, so that photochemical effects are limited to the outer shells of clouds.

The most energetic photons emitted by a bright star are retained within the HII region surrounding that star, and the energy cutoff corresponds to the ionisation potential of atomic hydrogen, 13.598 eV. Thus, in the neutral (HI) zone outside the HII region the interstellar medium is normally pervaded only by photons of energies up to that limit of 13.598 eV. This energy is lower than the ionisation potential of molecular hydrogen (15.4 eV), helium atoms (24.587 eV), oxygen atoms (13.618 eV), and nitrogen atoms (14.534 eV).

Figure 3.1. Top: The dark cloud Barnard 68 (B68) (Source: ESO). Bottom: The stellar jet of Herbig–Haro 49/50 as imaged by the Spitzer Telescope. (Courtesy of Bally, J., University of Colorado, JPL-Caltech, NASA.).

In general, therefore, these relatively abundant species are not photoionised and remain mainly neutral in diffuse clouds. Because reactions between these neutral atoms and molecular hydrogen are suppressed at the normally low temperatures of interstellar space (typically, less than or about 100 K), chemistry is not initiated among these species by starlight. However, although the radiation field cannot ionise H_2, it can dissociate the molecule via an electronic transition into one of the first two excited electronic levels, B and C, from which a cascade into the vibrational continuum of the ground state, X, is likely in a high fraction of excitations:

$$H_2(X, v'' = 0) + h\nu \rightarrow H_2(B, C; v') \rightarrow H_2(X; v'' = \text{continuum}) \rightarrow H + H$$

$$(3.1)$$

where $h\nu$ represents a photon of energy around 12 eV; and v'' and v' are the vibrational quantum numbers in the lower and upper electronic states. The H atoms from the dissociated H_2 molecule carry away a small amount of the energy that was contained in the UV photon.

So the H/H_2 balance depends on (and affects) the local radiation field and on the efficiency of H_2 formation on grain surfaces. The first major chemical change that occurs as radiation penetrates a mainly neutral cloud is the transition from a gas of mainly atomic hydrogen to one that is mainly molecular. The depth at which this transition occurs depends on the intensity of the external field and on the local gas number density; for diffuse clouds in the average interstellar radiation field, the transition occurs at a depth near $A_V \sim 0.1$ magnitudes, whereas in high-intensity and high-density regions the transition may be at a depth of several visual magnitudes.

However, other atomic species can be ionised by the local radiation field, and carbon (with an ionisation potential of 11.26 eV) and sulfur (10.36 eV) are the most important for driving the chemistry in the H_2-rich zone. The halogens fluorine and chlorine have different properties in the HI region. Fluorine (ionisation potential 17.42 eV) is neutral, whereas chlorine (12.97 eV) is ionised. Other relatively abundant elements, such as iron, silicon, and magnesium, can also be ionised by the interstellar radiation field, but these elements are probably almost entirely locked up in interstellar dust grains.

Carbon ions radiatively associate with molecular hydrogen

$$C^+ + H_2 \rightarrow CH_2^+ + h\nu \qquad (3.2)$$

where $h\nu$ signifies the emission of a photon. Reactions and associations of the product ion with H_2 molecules lead to CH_3^+ and CH_5^+. Dissociative recombinations of all these ions can form neutral C, CH, CH_2, and CH_4. These species may be a feedstock for further reactions; for example, the fast atom-radical reaction

$$CH + O \rightarrow CO + H \qquad (3.3)$$

can be a significant route to the very important tracer molecule carbon monoxide. Thus, in the region where H_2 becomes abundant the important element, carbon, is converted from C^+ to C and CO, and at a point where the UV field is sufficiently attenuated CO becomes a significant carrier of the available carbon. This occurs at depths typically of about two magnitudes in clouds in the average interstellar radiation field and somewhat deeper in situations of higher radiation intensity.

Similar exchange reactions involving nitrogen atoms can provide CN. The products of these schemes, CH, CN, and CO molecules, are indeed detected interstellar species (see Table 1.1). Sulfur ions, S^+, radiatively associate with H_2 molecules rather slowly, and the chemistry of sulfur-bearing species is initiated more efficiently in exchange reactions. For example, S^+ and S exchange reactions with CH or C_2 give CS, and reactions with OH lead to SO.

Chlorine ions react with H_2 to form HCl^+, which after several fast reactions ultimately becomes the neutral molecule HCl. Unusually for neutral atoms, fluorine atoms react directly with H_2 molecules to form HF. In diffuse clouds, HF is the dominant reservoir of fluorine.

Thus, photoionisations on their own lead to a characteristic but rather limited chemistry. At greater depths into the cloud, when the radiation field is more heavily extinguished, cosmic rays make a significant contribution to driving the chemistry, and the chemical results begin to approach those of dark cloud cosmic–ray driven chemistry (we discuss cosmic ray–driven chemistry in Section 3.3). See Figure 3.2 for a schematic diagram showing the main chemical structure of a photon-dominated region (a PDR).

In all PDRs, the radiation field is a major heat source and determines the local temperature. The temperatures in PDRs of high density and radiation field may reach values on the order of 10^3 K, whereas in diffuse clouds in the average interstellar medium the temperature is on the order of 100 K. In the region where hydrogen is molecular but other chemistry has not developed significantly, the important cooling lines are [OI] at 63 micron (for the brightest PDRs) and [CII] at 158 micron (for less bright PDRs). Important molecular coolants are ro-vibrational lines of H_2 and rotational lines of CO. These species are, consequently, important tracers of PDRs.

Although PDRs are most prominent near massive stars (where the radiation fields may be many times as intense as the mean interstellar radiation field), diffuse clouds are also PDRs created by the mean interstellar radiation field. If no other chemical driver is operating, the chemistry created by the interstellar radiation field is not extensive (see Figure 3.4a).

3.2.2 X-Ray–Dominated Regions

X-rays are emitted by a variety of objects, from active galaxies to massive stars. X-rays can also be drivers of chemistry when they affect on surrounding

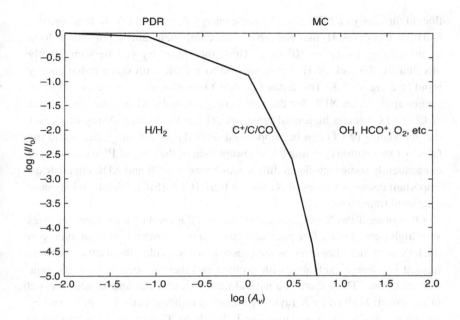

Figure 3.2. A schematic diagram of a PDR at an interface of an HII region with a molecular cloud (MC), showing the variation of the UV intensity with depth, and the depths into a cloud at which some important chemical transitions occur.

neutral matter. Soft X-rays (with photon energies less than about 1 keV) are absorbed close to the source, while hard X-rays (with photon energies greater than about 1 keV) – such as those generated near black holes in AGNs – have small absorption cross sections and may therefore affect large volumes of gas.

The chemistry initiated by X-rays may differ in some respects from that in PDRs because the X-rays are capable of ionising all material, including atomic and molecular hydrogen, helium, and CO. Indeed, CO^+ emission has been detected in strong X-ray sources. X-ray ionisation may lead to multiple ionisation of heavy elements in the Auger process. Most of the ionisation is caused by the secondary electrons released in this process. These secondary electrons ionise atomic and molecular hydrogen, and H^+ and H_3^+ become the dominant ions, so multiply charged heavy ions play only a small part in the chemistry. The H^+ and H_3^+ ions, as we see in Section 3.3, are also the ions created by cosmic ray ionisation. Thus, the effect of strong X-ray ionisation on the chemistry has some similarities to that driven by a high flux of cosmic rays. However, because high-energy X-rays penetrate the cloud more readily than stellar UV, the H/H_2 transition generally occurs much deeper into a cloud than in the case of a PDR with the same energy flux in the UV. In high-density and high-flux cases, H_2 may remain a minor component until depths into a cloud corresponding to $A_V \sim 10^3$ mag. Thus, the chemistry can be significantly affected by the lack of H_2, as compared to a PDR with comparable energy input (see Figure 3.3). The distinct $C^+/C/CO$ transitions observed in a PDR no longer apply in an XDR for the same energy flux. In XDRs, the abundances of C^+ and C remain high until large optical depths. In high–X-ray flux cases the abundance of CO can be suppressed until $A_V \sim 10^3$ magnitudes and the $C^+/C/CO$ transitions are much less abrupt than in the case of PDRs. There are consequently some significant differences between PDR and XDR chemistries. Important coolants/tracers of XDRs are [OI], [CII], [SiII], [FeII], and H_2 pure rotational transitions.

Of course, if the X-rays are so intense that the level of ionisation becomes very high (say, above a few percent relative to the number density of hydrogen nuclei), then the chemistry is suppressed as molecules themselves become ionised by charge exchange with protons and then dissociatively recombine with electrons. Thus, there is a natural limit to molecular abundances created in a chemistry driven by X-rays: the chemical richness cannot be enhanced by simply increasing the X-ray intensity indefinitely. This chemical sensitivity to X-ray intensity is discussed in more detail in Chapter 4.

3.3 Chemistry Initiated by Cosmic Rays

Cosmic rays are important drivers of interstellar chemistry. These fast particles, mostly hydrogen and helium nuclei, collide with and ionise atoms and molecules in the interstellar gas. They also heat the gas through the kinetic energy given to the ejected electrons. Theoretically, the most effective cosmic rays in ionising the gas are those with energies of a few MeV; unfortunately, the flux of these relatively low-energy cosmic rays is poorly determined at Earth because of modulation by the solar wind, so the actual interstellar cosmic ray ionisation rate in the Milky Way is still somewhat uncertain.

The importance of these relatively low-energy cosmic rays is that they can ionise atoms and molecules that are unaffected by UV starlight, including the most abundant species – hydrogen atoms and molecules and helium atoms:

$$H + cr \rightarrow H^+ + e + cr \qquad (3.4)$$

$$He + cr \rightarrow He^+ + e + cr \qquad (3.5)$$

$$H_2 + cr \rightarrow H_2^+ + e + cr \qquad (3.6)$$

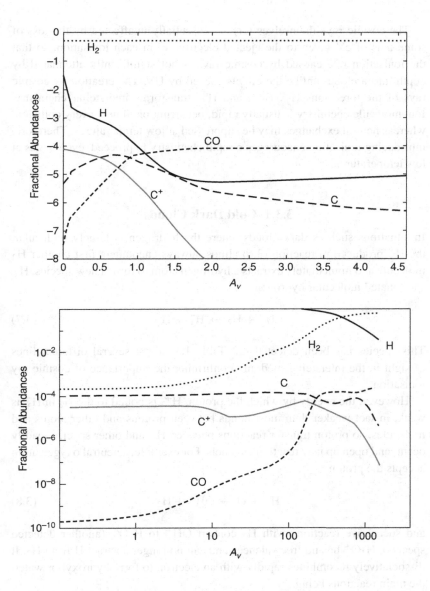

Figure 3.3. Fractional abundances of several species as a function of A_V for a PDR model (top) and an XDR model (bottom). Both models were run for a cloud of uniform density of H_2 molecules of 5×10^5 cm^{-3} and with a radiation flux in either UV or X-ray regimes of 0.274 erg s^{-1} cm^{-2}. The XDR curves are re-drawn from Meijerink, R. and Spaans, M., Astronomy & Astrophysics 436, 397 (2005).

The cosmic rays themselves (cr) are essentially unaffected by the loss of some tens of eV given to the ejected electron, e, in each ionisation, so that the ionisation rate caused by cosmic rays is not significantly attenuated by depth into a cloud, unlike the effects caused by UV. The creation by cosmic rays of the three ions H^+, He^+, and H_2^+, transforms interstellar chemistry. Ion–molecule chemistry is usually rapid, occurring on almost every collision, whereas neutral exchanges may be suppressed at low temperatures. Therefore, introducing ionisation encourages a rapid chemistry to proceed, even in gas at low temperatures.

3.3.1 Cold Dark Clouds

In situations such as dark clouds where the hydrogen is largely molecular, the H_2^+ produced in reaction (3.6) almost always encounters first another H_2 molecule and immediately extracts a hydrogen atom to form a new species, H_3^+ (protonated molecular hydrogen):

$$H_2^+ + H_2 \rightarrow H_3^+ + H \tag{3.7}$$

This species has been detected (cf. Table 1.1) along several different lines of sight in the interstellar medium, confirming the importance of cosmic ray ionisation.

However, the energy by which the proton, H^+, is bound to the H_2 in H_3^+ is weak, in fact weaker than most bonds between protons and other atoms and molecules, so proton transfer reactions between H_3^+ and other species readily occur, and open up new reaction channels. For example, a neutral oxygen atom accepts the proton

$$H_3^+ + O \rightarrow OH^+ + H_2 \tag{3.8}$$

and successive reactions with H_2 convert OH^+ to H_3O^+ (another detected species). H_3O^+ has no free valences and can no longer abstract H from H_2. It dissociatively recombines rapidly with an electron to form hydroxyl or water, the main reactions being

$$H_3O^+ + e \rightarrow OH + 2H, \tag{3.9}$$

and

$$H_3O^+ + e \rightarrow H_2O + H \tag{3.10}$$

Direct proton transfer from H_3^+ to CO creates the important tracer molecule HCO^+:

$$H_3^+ + CO \rightarrow H_2 + HCO^+ \tag{3.11}$$

The corresponding neutral species HCO can be created from HCO^+ in charge exchange reactions with low ionisation potential atoms, and also in a variety of other ways.

Although ion–molecule chemistry is clearly important in forming new molecules, it may also limit their abundances. Because helium has such a large ionisation potential (24.58 eV) the ion He^+ created by cosmic rays tends to destroy molecules through charge transfer, which – in this energetic transfer – may be dissociative. For example, CO undergoes charge transfer with He^+, and is dissociated:

$$CO + He^+ \rightarrow C^+ + O + He \tag{3.12}$$

This provides a source of C^+ that may then redistribute carbon from CO to other carbon-bearing species. This process of dissociative charge exchange with He^+ applies to all molecules formed in interstellar ion-molecule chemistry. Thus, He^+ provides an important limitation to the abundance of species arising in the chemistry initiated by cosmic rays.

Carbon atoms, like oxygen atoms, also accept a proton from H_3^+; they then form CH^+ molecules. Successive hydrogen abstraction reactions with H_2 molecules then rapidly create CH_2^+ and CH_3^+, and – more slowly – CH_5^+. Neutral hydrocarbons CH, CH_2, and CH_4 arise when these ions recombine with electrons.

The preceding examples show that cosmic rays provide an entry into the chemistry of oxygen-bearing and carbon-bearing species through proton transfer reactions with H_3^+. What about the chemistry of other elements? One might expect the chemistry of nitrogen to be similar to that of oxygen, but the reaction of nitrogen atoms with H_3^+ is believed to be suppressed at low temperature. N^+ ions could be created by direct cosmic ray ionisation of N_2 or other nitrogen-bearing molecules, but the reaction of N^+ with H_2 is also suppressed at low temperature. Evidently, nitrogen does not follow the pattern of oxygen; neither does sulfur.

However, by creating the first products of oxygen and carbon chemistry, cosmic rays open the door to a more complex chemistry, because these first products may interact with atoms and ions to form new products. For example, exchange reactions of OH with O, C, N, and S create O_2, CO, NO, and SO, respectively. Exchange reactions of CH with the same set of atoms create CO, C_2, CN, and CS.

Higher levels of chemical complexity can be attained by reactions involving these second-rank products, either among themselves or with existing atoms and ions, to form third-rank products. For example, SO and OH react to form SO_2. In another example, complex hydrocarbons arise from reactions of carbon ions with simpler hydrocarbons:

$$C^+ + CH_4 \rightarrow C_2H_3^+ + H \tag{3.13}$$

where the product $C_2H_3^+$ recombines with electrons to form the important molecule acetylene, C_2H_2, widely observed in interstellar and circumstellar environments. It is also a key molecule in the formation of the cyanopolyyne sequence, $HC_{2n}CN$ ($n = 1, 2, 3 \ldots$) initiated by the reaction of CN with C_2H_2.

In summary, the chemistry initiated by cosmic rays begins with the conversion of oxygen and carbon to several of their hydrides, both neutral and ionic (see Figure 3.4). Then other elements such as nitrogen and sulfur can be brought into play and a variety of products formed from simple exchange reactions. Greater complexity then arises when these products take place in a further stage of reaction. Molecules are being continually formed; they are also continually destroyed as they themselves are used up as reactants in synthesising new species. At the same time, destructive reactions such as those with He^+ tend to return material to simpler (atomic) components. The network of reactions may reach a chemical steady state if circumstances permit.

The lists of possible reactions involved in gas-phase interstellar and circumstellar chemistry can be large, extending to some thousands of reactions. Nevertheless, model calculations show that these schemes do not readily create significant abundances of molecules containing more than a few 'heavy' atoms (i.e., atoms other than hydrogen). The chemistry of larger species, such as ethanol (C_2H_5OH) and methyl formate (CH_3COOH), requires a different driver; we discuss this in Section 3.4.

In the picture developed here for dark cloud chemistry, it is clear that almost every time that a hydrogen molecule is ionised by a cosmic ray, ultimately a molecule involving oxygen, carbon, nitrogen, or sulfur is formed. We can therefore easily calculate from the cosmic ray ionisation rate the time required to convert most of the heavy atoms into molecules (t_{chem}). This time is on the order of 10^6 years for Milky Way parameters. This number scales with metallicity and inversely with the cosmic ray ionisation rate; we find that the chemical timescale, in years is:

$$t_{chem} \simeq 5 \times 10^6 \xi / \zeta \tag{3.14}$$

where ξ is the metallicity in units of the solar metallicity and ζ is the cosmic ray ionisation rate in units of 1×10^{-17} s^{-1}.

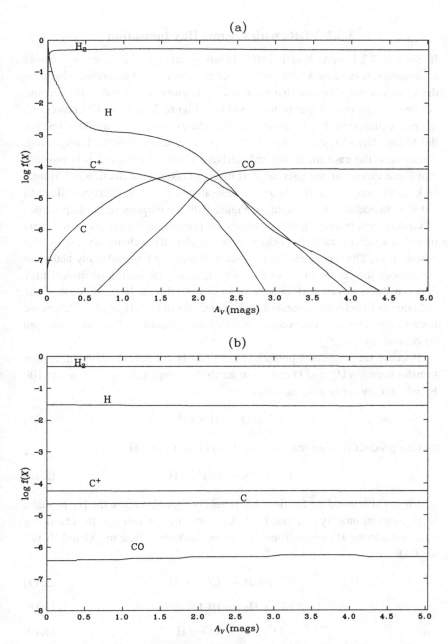

Figure 3.4. Fractional abundances, $f(X)$, of selected, X, species as a function of A_V into a 1-D semi-infinite cloud for models with (a) $10\times$ the mean interstellar radiation field intensity and no cosmic rays, and (b) no radiation field but $10\times$ the mean interstellar cosmic ray ionisation rate. The chemistry, as represented by CO, persists deep inside the cloud in case (b).

3.3.2 PDRs with Cosmic Ray Ionisation

In Section 3.2.1 we considered the chemistry arising when a gas of cosmic composition is irradiated by the interstellar radiation field, but no other chemical driver was acting. We saw that the resulting chemistry was rather limited and confined to the outer layer of the cloud (see Figure 3.4a; here, CO represents the consequence of the chemistry). In fact, the example is an unlikely one for the Milky Way Galaxy, where cosmic rays provide a significant background. This is also the case for active and starburst galaxies; however, it is possible that some quiescent red galaxies and dwarf galaxies (see Section 6.3.6) may lack significant cosmic ray fluxes. In general, however, cosmic rays will act as a driver, in addition to starlight. In Figure 3.4b we emphasize the importance of cosmic rays in driving the chemistry by presenting the results of a simple model in which the radiation field is switched off and the chemistry is driven by cosmic rays. The representative product molecule, CO, is uniformly abundant throughout the cloud. In general, however, this combination of drivers may occur at the edges of dark clouds, or in diffuse clouds. In these regions, the radiation field helps to maintain a high abundance of both H and H_2. Hence, we need to consider also the consequences of the ionisation of atomic hydrogen by cosmic rays.

Because the ionisation potentials of O and H are almost identical, charge transfer between H^+ and O can occur easily for temperatures above about 100 K (which may apply in these regions):

$$H^+ + O \rightarrow H + O^+ \tag{3.15}$$

and the product O^+ can react directly with H_2 to form OH^+:

$$O^+ + H_2 \rightarrow OH^+ + H \tag{3.16}$$

which (as discussed in Section 3.3.1) reacts successively with H_2 to make H_3O^+, and ultimately OH and H_2O. This provides an entry to the chemistry alternative to the H_3^+ route. There is also an alternative route to CO and HCO^+ via CO^+:

$$C^+ + OH \rightarrow CO^+ + H \tag{3.17}$$

$$CO^+ + H_2 \rightarrow HCO^+ + H \tag{3.18}$$

$$HCO^+ + e \rightarrow CO + H \tag{3.19}$$

Otherwise, the chemical network for these regions is similar to that for dark clouds where the chemistry is driven by cosmic rays. Of course, if the radiation fields in these regions are such that the hydrogen is almost entirely atomic, then the chemistry is very sparse indeed.

3.4 Chemistry and Dust

3.4.1 Surface Reactions on Dust Grains in Diffuse Clouds

The most abundant molecule in the Universe is H_2; in the Milky Way Galaxy H_2 is formed (almost entirely) in reactions that occur on the surface of dust grains. The surface of the dust grain acts as a catalyst in the reaction; it is not used up but is available for an unlimited number of reactions. The function of the dust grain is to trap one H atom long enough for a second H atom to arrive at the surface, locate the first, and react with it. The reaction

$$H + H - g \rightarrow H_2 + g \qquad (3.20)$$

(where $H - g$ represents an H atom weakly bound to a dust grain) releases about 4.5 eV per molecule. This is the difference between the dissociation energy of H_2 and the (rather small) energy of the bond between the H atom and grain surface. This energy of 4.5 eV is deposited partially in the grain as heat, partially as kinetic energy of the product H_2 molecule, and partly as energy of internal vibrational and rotational modes of the H_2 molecule.

These surface reactions were originally postulated to occur because of the absence of gas-phase mechanisms that are sufficiently fast to account for the observed H_2 abundances, but the concept as proposed has been confirmed by several elegant experiments in different laboratories using a variety of techniques, and supported by detailed quantum mechanical studies of the reaction. Experiments under appropriate physical conditions have been performed on different surfaces, representing different types of dust grain, and the reaction efficiency and energy budget have been shown to depend on the nature of the surface and on the gas and grain temperatures. However, the H_2 formation reaction appears to be sufficiently fast (compared to the H_2 loss rate discussed in Section 3.2.1) that under interstellar conditions in the Milky Way much of the hydrogen will be molecular. This requires that nearly all of the H-atoms that arrive on a grain surface must leave as part of a molecule of hydrogen. The newly formed H_2 molecule will typically have about 2 eV in internal modes and about 1 eV of kinetic energy. The remainder of the energy is deposited in the dust grain and will ultimately be radiated away at infrared wavelengths.

Of course, the H_2 formation process is not unopposed. In photon-dominated regions, H_2 can be destroyed by UV starlight of wavelengths near 100 nm,

as described in Section 3.2.1. At deeper positions within a cloud, where UV
starlight is heavily extinguished, H_2 is destroyed (rather slowly) by cosmic
rays. First, as we saw in Section 3.3, the H_2 is ionised and H_3^+ is created:

$$H_2 + cr \rightarrow H_2^+ + e + cr \qquad (3.21)$$

$$H_2^+ + H_2 \rightarrow H_3^+ + H \qquad (3.22)$$

but H_3^+ reacts quickly with the second most abundant molecule, CO,

$$H_3^+ + CO \rightarrow HCO^+ + H_2 \qquad (3.23)$$

which eventually recombines with an electron

$$HCO^+ + e \rightarrow CO + H \qquad (3.24)$$

so that the CO molecule is recovered. Thus, overall, the cosmic ray ionisation of
one H_2 molecule leads to the dissociation of one H_2 molecule into two H atoms.
For Milky Way conditions, the formation process is sufficiently effective that in
dark clouds very nearly all hydrogen is molecular, while even in diffuse clouds,
that is, PDRs with the mean interstellar radiation field, a significant fraction
of the hydrogen is molecular. For dark cloud conditions, where starlight is
excluded, the number density of hydrogen in atomic form is – in steady state –
given by

$$n(H) \simeq \zeta / R \qquad (3.25)$$

where ζ is the cosmic ray ionisation rate, and R is the H_2 formation rate
coefficient. For Milky Way conditions, $R \sim 10^{-17}$ cm^{3}s^{-1} and $\zeta \sim 10^{-17}$ s^{-1},
so that – in the steady-state regime – the number density of H atoms in a dark
cloud should be roughly 1 cm^{-3}.

The reaction to form H_2 is apparently very effective in diffuse clouds, so what
about the formation of other hydrides at dust grain surfaces? Here, supportive
laboratory evidence is largely absent, but some observational evidence suggests
that N atoms may be hydrogenated on the surfaces of dust grains. The nitrogen
hydrides NH, NH_2, and NH_3 have been detected in diffuse clouds (see Section
5.1.1). However, there is no consensus on their formation mechanisms. As
discussed in Section 3.3.1, nitrogen atoms or ions are inhibited from reacting
with H_2 in gas-phase reactions. However, if it is postulated that an efficient
hydrogenation reaction occurs on the surface of dust grains, then nitrogen
hydrides may be produced in the observed abundances. So the situation for NH
is currently similar to that for H_2 before the relevant laboratory studies were
made.

Surface reactions for carbon and oxygen atoms are generally not invoked in chemical models of diffuse clouds because gas-phase routes are available for these atoms (see Section 3.3.1). However, if one assumes that the hydrogenation of O and C occurs efficiently at dust grain surfaces and that these products (H_2O and CH_4) enter the gas-phase chemical network, then gas/grain chemical models predict molecular abundances that do not conflict with observations of molecules in diffuse clouds. Therefore, the jury is out on the question of whether N, O, and C atoms are hydrogenated at grain surfaces in diffuse clouds, until appropriate laboratory experiments provide a definitive answer.

3.4.2 Freeze-Out, Ice Formation, and Timescales

As discussed in the preceding section, it is as yet unclear whether surface reactions involving oxygen atoms contribute to chemistry of diffuse clouds. However, the detection of water ice mantles in cold dark interstellar clouds makes it inescapable that surface reactions on dust grains involving oxygen atoms create water molecules, and that a significant fraction of these H_2O molecules – possibly all of them – are retained at the surface to form ice. In cold clouds, gas-phase processes are simply too slow to account for the formation of ice mantles that contain a significant amount of the available oxygen, within the lifetime of the cloud.

Other species may simply stick to grain surfaces without further reaction. For example, CO will stick efficiently to surfaces at temperatures below about 25 K and is abundant in the ices that accumulate on grain surfaces. Some of the CO appears to be converted to CO_2 or other species. Other molecules produced by cosmic ray driven chemistry will also stick efficiently on low-temperature grains, that is, at 10–20 K, though the lightest species, H_2 and He, do not stick at these temperatures.

Thus, the ices that accumulate on dust grain surfaces are of a mixed composition, reflecting the local conditions and evolutionary history. Table 3.1 shows the average ice mantle composition for a number of objects in the Milky Way Galaxy, as determined by infrared spectroscopy. The relatively high abundance of CO_2, CH_3OH, and H_2CO in the ice (relative to H_2O) suggests that some processing of CO to these products is occurring, promoted by cosmic rays or by photons generated by cosmic rays in the cloud interior. Formaldehyde (H_2CO) and methanol (CH_3OH) appear to be stages in the surface hydrogenation of CO, and this conclusion is supported by laboratory experiments. Similarly, CO_2 may be the result of oxygenation of CO in reactions such as

$$CO + OH \rightarrow CO_2 + H \tag{3.26}$$

In many dark molecular clouds, the ices contain a significant fraction of the elements O, C, N, and S. For example, in many prestellar cores more than 90% of available carbon is in CO and CO_2 in ices. If there are no efficient processes returning material from the ice to the gas phase, then the gas phase is impoverished in the major elements other than hydrogen and helium, and gas-phase chemistry is severely inhibited. Thus, in the absence of a return mechanism, there is a natural time limit to gas-phase chemistry determined by the time for appreciable freeze-out to occur.

For CO molecules in molecular clouds of the Milky Way, the freeze-out timescale (in years) is:

$$t_{\rm fo} \simeq 10^6 (10/T_K)^{1/2} (10^4 {\rm cm}^{-3}/n_{\rm H}) \tag{3.27}$$

where the kinetic temperature in the gas is T_K and the number density of hydrogen (in all forms) is $n_{\rm H}$. This timescale depends also on the size distribution of the grain population; in equation 3.27 we have assumed a population as described in Section 1.3.2. Assuming that the size distribution is the same in all galaxies, this timescale scales with the dust:gas ratio. For the Milky Way, $t_{\rm fo}$ is on the same order of magnitude as the time for cosmic rays to drive a chemistry that converts an appreciable amount of the heavy elements as atoms into molecules (see Section 3.3.1). Freeze-out can remove significant amounts of material from the gas phase; see Figure 3.5.

Ices can be removed when the local temperature is raised towards $\sim$100 K, say, near a newly formed star. When the grains are warmed, the ices evaporate and the molecules they contained are returned to the gas. There are also some processes that can help to return molecules from solid to gas phase even at low temperature. However, these mechanisms may be dominated by the freeze-out. If so, then in some dense cores in dark clouds in the Milky Way only on the order of a few percent of gas remains in the gas phase.

3.4.3 Complex Chemistry in Interstellar Ices

A more complex solid-state chemistry appears to take place within the ice in some regions. In the vicinity of massive star formation, some very small cores ($\lesssim$0.1 pc) are found to contain a remarkable chemical complexity. The detected molecules are relatively large compared to those in dark clouds, and the regions are relatively very dense and warm. These small, dense, and warm regions are called *hot cores* and are found where star formation is occurring. They contain a variety of species, including the following relatively large molecules: acetic acid (CH_3COOH), acetone ((CH_3)$_2$CO), methyl cyanide (CH_3CH_2CN), methyl formate ($HCOOCH_3$), and ethanol (CH_3CH_2OH).

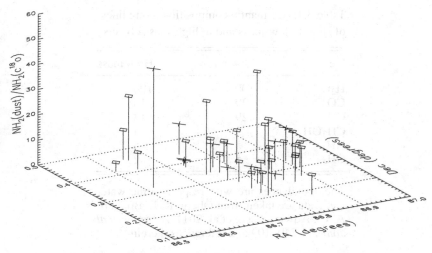

Figure 3.5. Depletion factor versus position in cloud for NGC 2071. Squares represent starless cores and crosses protostellar cores. The depletion factor is defined as the ratio of the hydrogen column density calculated from dust measurements to the hydrogen column density calculated from $C^{18}O$ observations. The more CO is depleted on to dust grains, the lower the hydrogen column density derived from the CO gas phase emission and hence the higher the value on the y-axis. (Reproduced, with permission from Christie, H., Viti, S., Yates, J. A., et al. 2012. *Monthly Notices of the Royal Astronomical Society*, 422, 968.)

Chemistry driven by cosmic rays cannot provide these relatively large species in the abundances required in hot cores (typically 10^{-8} relative to hydrogen) and another formation mechanism must be operating. It is generally assumed that these larger species arise in solid-state reactions in the ices that have formed on dust grain surfaces during the cold phase of cloud collapse, before they are evaporated into the gas phase when nearby star formation warms the gas and dust. The solid-state chemistry is promoted in the simple ices of H_2O, CO, CO_2, CH_4, and other species (see Table 3.1), by cosmic rays and by cosmic ray–induced photons. The injected energy creates radical and ions in the ices, and if these are sufficiently mobile in the ice then a new chemistry arises in which more complex species arise from the relatively simple ice. Laboratory work has examined some of the chemical pathways, and extensive computational modelling of these processes, involving many potential reactions, supports the view that chemical complexity may arise in this way (see Table 3.2). Models suggest that the variety and relative abundances of species created in this solid-state chemistry are consistent with the chemistry observed in regions of massive star formation.

Table 3.1. Ice mantle composition along lines
of sight to low-mass and to high-mass objects

Ice	Low mass	High mass
H_2O	100	100
CO	29	13
CO_2	29	13
CH_3OH	3	4
NH_3	5	5
CH_4	5	2

The abundances are with respect to the water
ice. Data are taken from Öberg, K. I., Boogert,
A. C., Adwin, A. C., et al. 2011. *Astrophysical
Journal*, 740, 109 and references therein.

3.5 Chemistry Initiated by Gas Dynamics

3.5.1 Shocks

A shock arises when an external event – such as a collision of one interstellar
cloud on another, or the impact of a rapidly expanding HII region on a nearby
cloud of cold neutral gas – drives a perturbation faster than the local sound
speed. The consequence is that kinetic energy of bulk motion is converted to
internal thermal energy, and the shocked gas is heated and compressed.

Table 3.2. Examples of radical–radical reactions on grains

$E_2(K)$	225	557	588	800	1189	1250	1425	1978	2254
Radical	H	CO	CH_3	HCO	NH	CH_3O	OH	NH_2	CH_2OH
H	H_2								
CO	HCO	x							
CH_3	CH_4	CH_3CO	C_2H_6						
HCO	H_2CO	x	CH_3CHO	OHCCHO					
NH	NH_2	HNCO	CH_2NH	HNCHO	N_2H_2				
CH_3O	CH_3OH	CH_3OCO	CH_3OCH_3	$HCOOCH_3$	CH_3ONH	$(CH_3O)_2$			
OH	H_2O	COOH (CO_2 + H)	CH_3OH	HCOOH	HNOH	CH_3OOH	H_2O_2		
NH_2	NH_3	NH_2CO	CH_3NH_2	NH_2CHO	$HNNH_2$	CH_3ONH_2	NH_2OH	$(NH_2)_2$	
CH_2OH	CH_3OH	$CH_2(OH)CO$	C_2H_5OH	$CH_2(OH)CHO$	$CH_2(OH)NH$	CH_3OCH_2OH	$CH_2(OH)_2$	$CH_2(OH)NH_2$	$(CH_2OH)_2$
$E_2(K)$	2500	1500 (80)							

The product of each reaction is shown in the box corresponding to the pair of
reactants. Data taken from Garrod, R. T., Weaver, S. L. Widicus, and Herbst,
E. 2008. *Astrophysical Journal*, 682, 283.

In the simplest case, the energy conversion at the shock wave is abrupt compared to other timescales in the gas, and the temperature and density jumps are effectively discrete. This is called a J shock. In the post-shock gas, radiative cooling begins to cool the gas but there is a finite period in which the gas is at an elevated temperature. For a completely molecular gas undergoing a shock of velocity v_s the immediate post-shock temperature is

$$T_{ps} \sim 5 \times 10^3 (v_s/10)^2 \qquad (3.28)$$

where v_s is in km s^{-1} and T_{ps} is in K. The cooling timescale depends on the chemical nature of the post-shock gas, but may be on the order of a thousand years for shocks of modest velocity.

More generally, the shock structure may be modified by the action of a local magnetic field on ions in the gas. In this case, the ions drift through the mainly neutral gas, depositing energy in the gas over a much wider range than in the simple J shock case. This more continuous effect is called a C shock. For the low ionisation expected in molecular clouds, and for typical magnetic fields strengths on the order of a microgauss, shocks with velocities less than about 40 km s^{-1} are likely to be C shocks. Because the energy in a C shock is dumped over a larger physical extent than in a J shock, the resulting temperatures are generally lower, typically $\sim 10^3$ K for a shock speed of ~ 30 km s^{-1}, than in a J shock, see Figure 3.6.

Such dynamic events are common in interstellar and circumstellar media, and may be detected spectroscopically through indicators such as anomalous quantum level populations in atoms and molecules. Depending on the shock speed and local conditions, the enhanced temperature may be very large, and in extreme cases the shocked gas may be collisionally ionised and molecules collisionally dissociated. For a J shock in a molecular cloud with number density $n(H_2) \sim 10^4$ cm^{-3}, molecular hydrogen would be dissociated in a shock with speed ~ 25 km s^{-1} or larger. In a C shock, by contrast, the shock speed at which H_2 dissociation becomes important rises to ~ 45 km s^{-1}, because the post-shock temperatures are lower in C shocks than in J shocks.

These critical velocities at which H_2 can be collisionally dissociated are important. Above those velocities the shocked gas contains abundant hot atomic hydrogen that is capable of destroying all types of molecules, for example:

$$H + H_2O \rightarrow OH + 2H \qquad (3.29)$$

and

$$H + OH \rightarrow O + 2H \qquad (3.30)$$

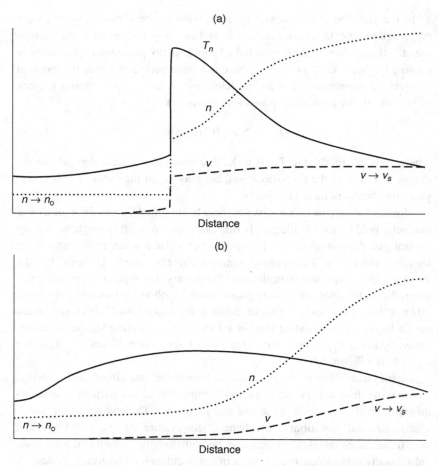

Figure 3.6. A schematic diagram, not to scale, of the density and temperature structure across a J shock (a) and a C shock (b).

Thus, these shock velocities are the limits above which chemical complexities begin to decrease. If the post-shock temperature rise is modest, say to a few thousand Kelvin, then a characteristic shock chemistry may arise; for example, reactions that may lead to the production of water. The reactions

$$O + H_2 \rightarrow OH + H \tag{3.31}$$

and

$$OH + H_2 \rightarrow H_2O + H \tag{3.32}$$

which are suppressed at the low temperatures (~ 10 K) of interstellar clouds by activation energy barriers of a few tenths of an eV in the reaction path, become rapid at temperatures $\sim 10^3$ K. Thus, in an interstellar shock, the chemical route to form water involves two of the most abundant components of the interstellar gas, whereas (as we have seen in Section 3.3.1) at low temperature the chemistry to form water is initiated by the minor ions H^+ and H_3^+ produced by cosmic ray ionisation. The consequence is that although gaseous H_2O is always a minor species in cold clouds, in shocked gas it may be abundant and can contain most of the oxygen that is not already bound in CO.

Many neutral–neutral reactions, and some ion–molecule reactions, are impeded by activation energy barriers of a magnitude similar to those in the oxygen–hydrogen system. For example, atomic sulfur behaves in a similar way to oxygen, and the detection of a rich sulfur chemistry can be taken as an indicator of the presence of a shock. In cold gas, carbon chemistry is initiated by the slow radiative association of C^+ with H_2 or by proton transfer to C atoms from H_3^+. In modest shocks, however, the reaction

$$C^+ + H_2 \rightarrow CH^+ + H \tag{3.33}$$

(which is impeded at low temperatures by an activation energy barrier of about 0.4 eV) provides a fast entry into the carbon chemistry, involving, as it does, two of the most abundant species in the interstellar gas.

In general, low-velocity shocks enable atomic species to react directly with molecular hydrogen. Although the hydrides formed may then proceed to react further, hydride molecules such as H_2O and H_2S are the immediate signatures of low-velocity shocks.

3.5.2 Interfaces

The elevated temperatures in post-shock gas are clearly important in driving reactions that are impeded by activation energy barriers at low temperatures. However, a temperature rise in cool molecular gas can also be created by mixing hot gas with the molecular gas. Such a situation arises in the turbulent interface that exists when a stellar wind erodes the molecular gas in the dense core surrounding a young star.

Chemically, this situation is somewhat different from a shock, because the wind gas is not only hot but also ionised. The mixing process both warms the gas and raises the level of ionisation in the turbulent interface, thereby enhancing ion–molecule chemistry in the gas. The process should continue

Table 3.3. Observed
fractionation for a sample of
molecule towards the low-mass
protostar IRAS 16293-2422

HDO/H_2O	0.03
$HDCO/H_2CO$	0.15
D_2CO/H_2CO	0.05
CH_2DOH/CH_3OH	0.9
CH_3OD/CH_3OH	0.04
CHD_2OH/CH_3OH	0.2

until the stellar wind has completely eroded the dense core in which the star was formed, a process that may take about a million years.

In the wind/core situation, it is likely that the gas in the core has been dense and cold enough for icy mantles to have been deposited on the grains (see Section 3.4). If so, then any molecules that were in the ice will be evaporated in the warm turbulent interface and will take part in gas-phase ion–molecule reactions.

3.6 Isotopes in Interstellar Chemistry

Deuterium is present in the Milky Way Galaxy at an abundance, on average, of 1.4×10^{-5} relative to hydrogen. Naïvely, one would expect that the ratio of deuterium-containing molecules, such as NH_2D, to their hydrogen-containing version, for example, NH_3, will be a similarly small number, and that the relative abundance of more strongly deuterated species, such as ND_3, would be vanishingly small. In fact, this is a situation in which intuition leads to an incorrect conclusion. There are locations in which deuterated species are many orders of magnitude larger than expected (see Table 3.3). The effect is not limited to deuterium (though is most pronounced for that species); the isotope ^{13}C is also found to be enhanced over ^{12}C in some circumstances. This enhancement of a rare isotope in a molecule is called *fractionation*, and arises in isotope exchange reactions in ion–molecule chemistry. We discuss fractionation here in terms of deuterium chemistry. The varieties of a molecule that differ only in their isotopic composition are called *isotopologues*.

In dark interstellar clouds, deuterium is largely locked in deuterium hydride, HD, by the reaction

$$D^+ + H_2 \rightarrow H^+ + HD \tag{3.34}$$

where the D^+ is formed by charge exchange with H^+

$$D + H^+ \rightarrow D^+ + H \tag{3.35}$$

and the protons are a result of cosmic ray ionisation. Reactions of HD with molecular ions containing hydrogen can exchange deuterium for hydrogen:

$$XH^+ + HD \rightarrow XD^+ + H \tag{3.36}$$

because D is very slightly more strongly bound to X than H. Although the potential well is the same in both cases, XH^+ and XD^+, the ground vibrational level of the heavier isotope sits lower in the potential well than that of the lighter isotope. The energy difference is typically equivalent to a few hundred degrees Kelvin. Therefore, at temperatures below that level, the reaction (3.36) preferentially runs in the forward direction, as shown, and deuterium is transferred from the main reservoir, HD, into XD^+. However, at temperatures above that level the reaction can proceed easily forward and backwards, so fractionation is not enhanced. Of course, all XD^+ species preferentially formed at low temperatures can undergo a variety of reactions, and so distribute the deuterium widely among the chemistry. A key species in this distribution is the isotopologue of H_3^+, which we have seen is important in dark cloud gas-phase chemistry; the isotopologue is H_2D^+, which is more stable than H_3^+ by an energy difference equivalent to 178 K:

$$H_3^+ + D \rightarrow H_2D^+ + H \tag{3.37}$$

or

$$H_3^+ + HD \rightarrow H_2D^+ + H_2 \tag{3.38}$$

The reaction (3.38) is in fact a more effective route to H_2D^+ at low temperatures. H_2D^+ can be particularly enhanced when CO is highly depleted as CO is one of the major destroyers of H_2D^+.

The hydrogen-related isotopologues of ammonia, NH_3, are NH_2D, NHD_2, and ND_3. The deuterium fractionation for these species can be very high in cold gas. These fractionations are created by successive reactions of the type

$$NH_3 + H_2D^+ \rightarrow NH_3D^+ + H_2 \tag{3.39}$$

$$NH_3D^+ + e \rightarrow NH_2D + H \tag{3.40}$$

Table 3.4. Examples of tracers of interstellar processes under selected
physical conditions

Driver	Region	Tracers
UV radiation	PDRs	CS, H_2O, CO_2, OCS, ...
X-rays	XDRs	HNC, HCO^+, ...
Cosmic rays	Dark clouds	C_2H, H_2O^+, H_3^+, CN, HCN, HNC, ...
Dust	Hot cores	CH_3CN, CH_3OH, NH_3, SO_2, ...
Dynamics	Shocks, interfaces	SiO, H_2O, CH_3OH, HCO^+, ...

Similar fractionations can occur for other elements; however, the relative mass
difference between D and H is larger than for any other pair of isotopes. Thus,
the energy difference is smaller for all other isotopologues. For example, in the
reaction

$$^{13}C^+ + {}^{12}CO \rightarrow {}^{12}C^+ + {}^{13}CO \tag{3.41}$$

the ^{13}CO molecule is more stable than ^{12}CO merely by 35 K.

The ability to form H-related isotopologues by these schemes of low temperature interstellar chemistry increases the number of detected molecular species
enormously.

3.7 Conclusions

Even this brief introduction to interstellar chemistry can tell us something about
molecules that could be used as tracers of various kinds of region, assuming that one driver plays a dominant role. We summarise this information in
Table 3.4.

3.8 Further Reading

Cernicharo, J., and Bachiller, R. eds. 2011. *The Molecular Universe IAU Symposium 280*. Cambridge: Cambridge University Press.

Draine, Bruce T. 2011. *Physics of the Interstellar and Intergalactic Medium*. Princeton Series in Astrophysics. Princeton, NJ: Princeton University Press.

Duley, W. W. and Williams, D. A. 1984. *Interstellar Chemistry*. San Diego: Academic Press.

Hartquist, T. W. and Williams, D. A. eds. 1998. *The Molecular Astrophysics of Stars and Galaxies*. New York: Oxford University.

Lemaire, J. L., Vidali, G., Baouche, S., Chehrouri, M., Chaabouni, H., and Mokrane, H. 2010. Competing mechanisms of molecular hydrogen formation in conditions relevant to the interstellar medium. *Astrophysical Journal*, 725, L156.

Tielens, A. G. G. M. 2005. *The Physics and Chemistry of the Interstellar Medium*. Cambridge: Cambridge University Press.

Whittet, D. C. B. 2003. *Dust in the Galactic Environment*. Bristol, U.K.: Institute of Physics Publishing.

4

Physical Processes in Different Astronomical Environments

In Chapter 3 we saw that the chemistry producing molecular tracers in interstellar space is driven in a variety of ways: by electromagnetic radiation at UV and X-ray wavelengths; by ionisation caused by cosmic ray particles; by reactions on grains or in their icy mantles; and in chemistry induced by gas dynamics. In many types of region, some or all of these processes may act together; in others, one of them may dominate. The atomic and molecular tracers produced reflect the nature of the chemical drivers that are operating.

However, there is another complication. Each chemical driver does not always operate at the same rate; a driver may vary from place to place in the interstellar medium of a galaxy, or from galaxy to galaxy, or even from time to time. For example, the local UV radiation field in a galaxy with a transient high star-formation rate may be much more intense than the value corresponding to the mean intensity in the Milky Way Galaxy. Similarly, cosmic rays are accelerated by magnetohydrodynamical events and so locations of high dynamical activity may have much greater fluxes of cosmic rays than in more quiescent regions. Also, chemistry on dust grains is obviously affected by the dust:gas ratio and by the nature of the surfaces of the dust; both of these may vary from place to place within the Milky Way Galaxy, and from galaxy to galaxy. Finally, gas dynamical events causing shocks or turbulent mixing can be an important driver of the chemistry. Here, the key parameters are the velocity of the shocks and the extent of the mixing of two gases at an interface. These depend on the local conditions and events. Thus, all the four driver mechanisms that we have described may vary from place to place within a galaxy, or from one galaxy to another, or indeed they may vary in time. In this chapter, we consider the effects on the chemistry produced by a driver, allowing each of the chemical drivers to vary in activity.

4.1 Varying the Intensity of Electromagnetic Radiation

4.1.1 Photon-Dominated Regions

In Section 3.2.1, and illustrated in Figure 3.2, we saw that a photon-dominated region (PDR) has a characteristic structure that is set up as the optical and UV radiation from a massive star (or cluster of stars) penetrates a cloud and is absorbed by dust and gas.

First, there is an H/H$_2$ interface beyond which the hydrogen is mainly molecular while the rest of the gas remains mainly atomic. In this zone, the main coolants/tracers are [OI] and [CII]. Then, at greater depths into the cloud, a C$^+$/C/CO transition occurs, beyond which a significant fraction of the carbon is in CO. In this zone, rotational lines of CO are prominent tracers. Beyond that zone, exchange reactions between oxygen and nitrogen atoms and simple hydrocarbons lead to the formation of species such as CN, NH, and O$_2$. Deeper within the cloud, the external radiation field is more heavily extinguished and the rate of chemistry driven by it declines rapidly. The main driver of chemistry in these deeper regions is ionisation caused by cosmic rays (Section 3.3); cosmic ray fluxes have no significant depth dependence, so neither does the chemistry driven by cosmic rays.

For a dark cloud in the average radiation field of the Milky Way, the H/H$_2$ transition occurs at a depth corresponding to A$_V \sim 0.1$ mag, while the transition to CO occurs at $A_V \gtrsim 2$ mag (see Figures 3.2 and 3.3 and Section 3.2.1). What happens to this PDR structure when the intensity of the external radiation field is much larger? Obviously the intensity may be very much larger close to massive stars or in galaxies with high star-formation rates.

In crude terms, the structure remains the same but is moved deeper into the cloud. The region dominated by atomic hydrogen extends farther, and the H/H$_2$ transition occurs deeper into the cloud. Similarly, the transition from carbon atoms and ions to CO also occurs deeper within the cloud. These effects occur because the external radiation field tends to dissociate H$_2$ and CO molecules and photoionise C atoms. When the external field is more intense, then for a given density, the position at which reactions forming H$_2$ and CO can compete with the photoprocesses moves to a greater depth where interstellar extinction causes the intensity to be weaker. Because formation and recombination reactions scale as the square of the density, an increase in density moves the transitions to lesser depths in the cloud. Figure 4.1 shows how increasing the external radiation field intensity shifts the transitions to greater depths, for two number densities.

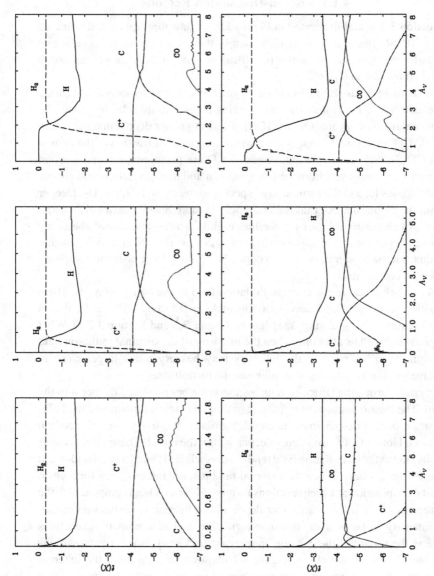

Figure 4.1. Fractional abundances, $f(X)$, of species $X \equiv H$, H_2, C^+, C, CO as a function of A_V for increasing radiation fields (left: mean interstellar field; centre: $100\times$ mean interstellar field; right: $10^4 \times$ mean interstellar field) for two hydrogen number densities, $n_H = 100 \ \mathrm{cm}^{-3}$ (top panels) and $n_H = 10^4 \ \mathrm{cm}^{-3}$ (bottom panels).

4.1.2 X-ray–Dominated Regions

The H/H_2 transition in X-ray–dominated regions (XDRs) depends on both the number density in the cloud and the flux of energy in X-rays impinging on it. Higher densities drive faster H_2 formation, so the H/H_2 transitions move towards the cloud edge when the cloud density increases. Higher X-ray fluxes, on the other hand, dissociate H_2 molecules more quickly, and tend to push the transition deeper into the cloud.

Using H_2 formation rates appropriate for the Milky Way, the results of detailed modelling show that for rather low number densities, say $n_H \sim 1000$ cm^{-3}, an increase in the X-ray flux from a value equivalent to one thousand times the mean UV energy flux in the Milky Way to 100 times this value moves the transition from $A_V \sim 1000$ mag to around five times deeper.

At higher number densities, say 3×10^5 cm^{-3}, the low X-ray flux is insufficient to dissociate the H_2 and the hydrogen is entirely molecular. However, the higher X-ray flux pushes the H/H_2 transition to $A_V \sim 600$ mag (see Figure 3.3).

As noted in Section 3.2.2, the distinct C^+/C/CO transitions observed in PDRs do not exist in XDRs because X-rays penetrate the cloud more readily than UV rays. Hence, C^+ and C are abundant to high depths into the cloud. When the X-ray flux is very high, then the CO abundance is generally low until depths corresponding to $A_V \sim 10^3$ mag. This is also the case even when the X-ray flux is low, except when the number density is high enough ($n_H \sim 10^5$ cm^{-3}) so that the formation of CO dominates the loss; in the latter case, CO is abundant throughout the cloud (see Figure 3.3).

In general, chemistry is suppressed in XDRs as compared with PDRs, so the important tracers are not molecular lines; they are forbidden atomic and ionic lines, such as [CII] at 158 μm, [OI] at 63 and 145 μm, [CI] at 610 μm, [SiII] at 18.7 and 33.5 μm, and [FeII] at 1.64 μm. Conventional dark cloud molecular species such as HCN, HNC, CS, HCO$^+$, and C_2H, for example, are generally at very low and undetectable levels unless at depths corresponding to $A_V \sim 1000$ mag. However, these species become more abundant and may be detectable in cases in which the density is very high and the X-ray flux is very low.

4.2 Varying the Cosmic Ray Ionisation Rate

As we have seen in Section 3.3, cosmic rays are important drivers of the chemistry, especially in the interiors of dark clouds where starlight is excluded. One might expect, therefore, that increasing the cosmic ray ionisation rate would increase the chemical abundances in the gas. In fact, this is not generally the case. Increasing the cosmic ray ionisation rate does of course increase the rates of formation of many molecules, but it also increases their rates of loss through other chemical reactions, and the balance between formation and destruction is rather subtle. In general, molecular abundances initially remain broadly the same if the cosmic ray ionisation rate is initially increased by modest amounts.

However, if the cosmic ray ionisation rate is very significantly increased, then H_2 is destroyed at such a rate that H_2 formation cannot compensate. Because the chemistry of almost all molecular species depends on H_2, the decline in the H_2 abundance means that all molecular species also decline. Figure 4.2 shows this result for gas irradiated by cosmic rays generating an ionisation rate between 10^{-18} and 10^{-12} s^{-1}. In this calculation, the gas is assumed to have metallicity of 0.1 of that of the Milky Way and a hydrogen number density of 10^4 cm^{-3}. Evidently, Figure 4.2 shows that the cosmic ray ionisation rate in the Milky Way Galaxy ($\sim 10^{-17}$ s^{-1}) is almost optimum for driving interstellar chemistry.

4.3 Varying the Dust: Gas Ratio and the Metallicity

We have argued in Section 3.4 that dust grains influence the chemistry in interstellar clouds in several ways: they extinguish starlight and protect molecules in cloud interiors from photodissociation; they may catalyse the formation of molecules directly on their surfaces; and they also provide substrates on which ices may be deposited in denser regions of interstellar space. Solid-state chemistry in those ices may then give rise to molecules not readily produced in gas-phase reactions. Therefore, variations in the dust:gas mass ratio may be important in the chemistry.

The dust grains in external galaxies may be expected to be similarly active in surface chemistry as Milky Way dust grains. However, this conclusion is also dependent on the size distribution. If, for example, the dust grains are predominantly smaller than those in the Milky Way, then – for the same dust:gas ratio – they would provide more surface area per unit volume and (if of a similar composition) be more chemically active than Milky Way dust. In the absence of clear evidence, it is often assumed that dust grains in different locations have the same size distribution as Milky Way dust. Then the catalytic effects of the dust (either in surface reactions such as H_2 formation, or in processing of icy mantles to produce large molecules such as methyl formate, CH_3COOH, or ethanol, C_2H_5OH) will simply be proportional to the dust:gas ratio.

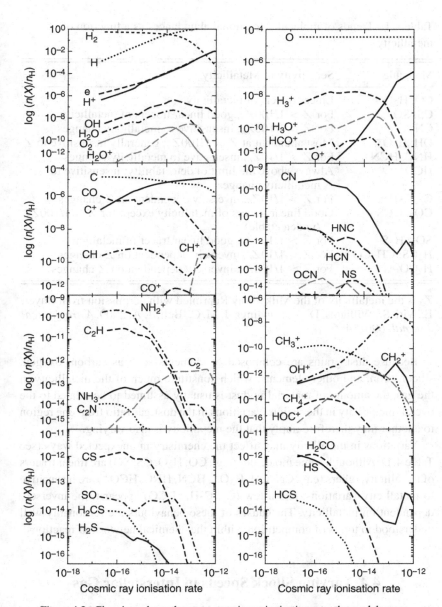

Figure 4.2. Chemistry dependence on cosmic ray ionisation rate; the models were run at a total hydrogen number density of 10^4 cm^{-3} and metallicity of 0.1 solar. The results are given at a depth into the cloud corresponding to $A_v = 3$ mag. (Reproduced, with permission from Bayet, E., Williams, D. A., Hartquist, T. W., and Viti, S. 2011. *Monthly Notices of Royal Astronomical Society*, 414, 1583.)

Table 4.1. Trends of molecular fractional abundances as a function of metallicity

Molecule	Sensitivity to Metallicity
CO, H_2O	Linear tracers of metallicity
CS, SO	For $Z > 1/100Z_o$, good linear tracers of metallicity
CN	For $Z > 1/100Z_o$, insensitive to metallicity changes
OH, H_3O^+	Most abundant at $Z = 1/100Z_o$; generally insensitive to Z
HNC, HCN	For $Z > 1/10Z_o$, insensitive to metallicity changes
HCO^+	Always above the limit of detectability; insensitive to metallicity changes
C_2, C_2H	For $Z > 1/10Z_o$, inversely dependent on Z changes
CO_2, OCS	Good linear tracers of metallicity except for $Z < 1/100Z_o$ (undetectable)
SO_2, H_2S	For $Z > 1/100Z_o$, good linear tracer of metallicity
H_2CS, CH_2CO	For $Z > 1/10Z_o$, inversely dependent on Z changes
H_2CO	For $Z > 1/100Z_o$, inversely dependent on Z changes

Z_o is the metallicity of the Milky Way Reprinted with permission from Bayet, E., Viti, S., Williams, D. A., Rawlings, J. M. C., Bell, T. A. 2009. *Astrophysical Journal*, 696, 1466.

Because dust grains are composed of elements such as carbon, oxygen, silicon, iron, and other elements, which constitute much of the metallicity in the gas, the amount of material in dust is usually assumed to be related to the overall metallicity in the region. Variations in the dust:gas ratio from one region to another may therefore imply that the metallicity is also varying.

Variations in metallicity may affect the chemistry in unexpected ways (see Table 4.1). Although some molecules (e.g., CO, H_2O, CS, SO) are linear tracers of metallicity, others (e.g., CN, OH, H_3O^+, HCN, HNC, HCO^+) are insensitive to metallicity variations, and a few (C_2, C_2H, H_2CO) appear to be inversely dependent on metallicity. The causes of these behaviours are reasonably well understood in terms of competition within the chemical network of reactions.

4.4 Varying Shock Speeds in Interstellar Gas

Shocks convert kinetic to thermal energy. If the gas temperature is raised from low values typical of molecular clouds ($\sim$10 K) to more than 10^3 K then, as discussed in Section 3.5, some reactions involving abundant species become efficient. In particular, water and some sulfur-bearing species are considered to be signatures of shocks.

However, as we have seen in Section 3.5.1, the range of shock speeds over which such tracer molecules are produced is rather limited. Once the shock speed and the post-shock temperature are high enough that much of H_2 is dissociated, then reactions of hot H atoms with all molecules remove them quickly. Shocks with speeds in this range are called dissociating shocks. For nonmagnetic shocks, the range over which shock signature molecules are produced is roughly 3–25 km s^{-1}. For magnetic shocks, the range extends up to about 45 km s^{-1}. Even in a dissociating shock, chemistry can also occur behind the shock as the gas cools, if some H_2 is available.

Post-shock temperatures $\sim 10^3$ K enable a rapid conversion of oxygen atoms, not already bound in CO, first to OH and then to H_2O. The ratio of abundances H_2O:OH can be a sensitive probe of density and radiation field in the shocked gas. When OH is abundant in dark clouds, reactions of O, C, and S with OH then form O_2, CO, and SO. At temperatures $\sim 10^3$ K reactions of C (in dark clouds) and C^+ (in diffuse clouds) initiate the formation of carbon hydrides. Somewhat higher post-shock temperatures are required to drive reactions of S and S^+ with H_2 to initiate sulfur chemistry in dark clouds. We list here a few examples of molecular shock tracers:

H_2O: As long as the shock velocity is greater than 10 km s^{-1} water is enhanced by shocks and, especially in a C-type shock, it does not dissociate easily (the main destruction route, via reaction with atomic hydrogen, has a very high barrier ~ 9000 K).

SiO, CH_3OH, and SO: These three species are among those routinely observed to be enhanced in energetic outflows; methanol formation in the gas phase is mainly via the reaction of CH_3^+ with H_2O so its large abundance in shocked environments may be mainly due to the enhanced abundance of water. However, SiO and SO are enhanced due to grain sputtering, rather than via high temperature gas-phase reactions, so although they do trace shocked environments they are an indirect product of the passage of the shock.

4.5 Timescales

It is clear from the preceding sections that changing the physical conditions (UV flux, cosmic ray ionisation rate, etc.) that determine the chemical drivers will also change the chemistry and the abundances of the molecules produced by those drivers. Although this book is primarily about identifying molecular tracers and how they can help to determine local physical conditions, it is worth restating here that molecules also have an active role in controlling some important aspects of the gas, in particular, its heating and cooling and its level of ionisation. Therefore, molecules are intimately concerned with the physical evolution of the gas, especially, for example, in star formation.

To illustrate this point we list in Table 4.2 the timescales of processes relevant to low-mass star formation in interstellar clouds. We met the chemical timescale and the freeze-out timescale in Chapter 3, but it is worth noting that there are other timescales of importance. Gravity determines the free-fall timescale, and this depends only on the total number density but not on the chemistry. The chemical timescale determines how soon an atomic gas becomes molecular; here we assume that the conversion is driven by cosmic rays. The freeze-out timescale determines when gas-phase molecular abundances are significantly reduced by ice formation on grain surfaces. The desorption timescale determines how quickly molecules may be desorbed from ices to the gas phase. The cooling timescale determines how quickly the gravitational potential energy of the cloud can be radiated away by molecules, and the ambipolar diffusion timescale determines how quickly magnetic support for a cloud can be removed.

We show in Figure 4.3 how these timescales depend on density, metallicity, cosmic ray ionisation rate, and UV radiation flux in a cloud of $A_V = 10$ mag. When the timescales do not differ too much, then the implication is that low-mass star formation can occur. But if, for example, the cooling time is long compared with the free-fall time, the gravitational potential energy of a collapsing core cannot be radiated away sufficiently quickly and therefore star formation will not occur. Similarly, if the ambipolar diffusion timescale is too long, then magnetic support will remain significant and will impede the gravitational collapse. Note that these arguments apply to the formation of stars of low mass. The formation of stars of high mass is generally considered to be triggered by violent dynamical events, rather than the quiescent process implied by the equations in Table 4.2.

4.6 Conclusions

We present in Table 4.3 some conclusions regarding the response of some important molecular tracers when the chemical drivers producing those species vary from canonical Milky Way values. The significance of this table will be clearer when we look in Chapter 6 at the tracers produced in galaxies in which physical conditions (and the drivers) are very different from those in the Milky Way.

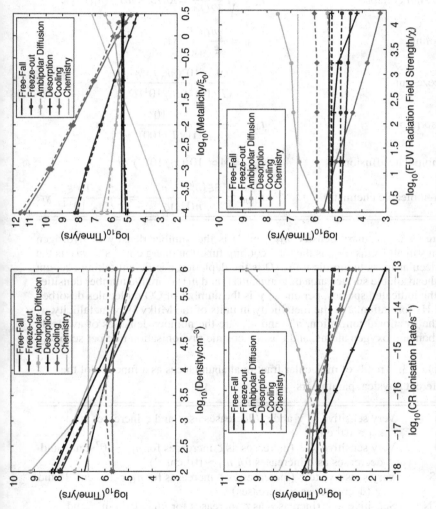

Figure 4.3. Variation of timescales with changes in density, metallicity, cosmic ray ionisation rate, and UV irradiation in a cloud of $A_V = 10$ mags. (Reproduced by permission of the AAS from Banerji, M., Viti, S., Williams, D. A., and Rawlings, J. M. C. 2009. *Astrophysical Journal* 692, 283.)

Table 4.2. Relevant timescales for interstellar processes

Timescale	Equation
Free-fall collapse	$t_{\mathrm{ff}} = \sqrt{\dfrac{3\pi}{32G\rho}} = 0.75 \times 10^8/(n_{\mathrm{H}})^{\frac{1}{2}}$ yr,
Cooling	$t_{\mathrm{cool}} = \dfrac{\frac{5}{2}n_{\mathrm{H}}kT}{\xi\Lambda_{\mathrm{tot}}}$ yr,
Freeze-out	$t_{\mathrm{fo}} = \dfrac{0.9 \times 10^6(m_X/28)^{\frac{1}{2}}}{(T/10)^{\frac{1}{2}}(n_{\mathrm{H}}/10^4)D}$ yr,
Desorption	$t_{\mathrm{des}} = \dfrac{1 \times 10^9}{\left[n_{\mathrm{H}}n(\mathrm{H})(T/100)^{\frac{1}{2}}\gamma\xi\right]}$ yr,
Ambipolar diffusion	$t_{\mathrm{amb}} = 4 \times 10^5(x_i/10^{-8})$ yr.
Ion–molecule chemistry	$t_{\mathrm{chem}} \simeq \dfrac{3\xi\left(n_{\mathrm{C}}^{\mathrm{tot}} + n_{\mathrm{O}}^{\mathrm{tot}}\right)}{n(\mathrm{H}_2)\zeta} = \dfrac{5 \times 10^6\xi}{\zeta/1 \times 10^{-17}}$ yr.

Here, ρ is the mass density, n_{H} (cm^{-3}) is the number density of hydrogen atoms in all forms, Λ_{tot} is the total cooling function in erg cm^{-3} s^{-1}, m_X is the molecular mass in atomic units, D is the depletion coefficient taking account of the available surface area of grains, $n(\mathrm{H})$ and $n(\mathrm{H}_2)$ are the number densities of the indicated species, per cm^{-3}, γ is the number of CO molecules desorbed per H$_2$ formation, ξ is the metallicity in units of the Milky Way metallicity, x_i is the fractional ionization, $n_{\mathrm{C}}^{\mathrm{tot}}$ and $n_{\mathrm{O}}^{\mathrm{tot}}$ are the number densities of available carbon and oxygen atoms, and ζ is the cosmic ray ionisation rate per second.

Table 4.3. Trends of molecular fractional abundances as a function of the different physical parameters

CS	Very sensitive to χ and ζ (decreases as χ and ζ increase) at $n_{\mathrm{H}} \sim 10^4$ cm^{-3}
SO	Very sensitive to ζ (increases as ζ increases for $n_{\mathrm{H}} \sim 10^7$ cm^{-3} and decreases as ζ increases for $n_{\mathrm{H}} \sim 10^4$ cm^{-3})
H$_2$S	Very sensitive to ζ (increases as ζ increases for $n_{\mathrm{H}} \sim 10^7$ cm^{-3}) and χ (decreases as χ increases)
HCN	Sensitive to ζ (increases as ζ increases for $n_{\mathrm{H}} \sim 10^7$ cm^{-3} and decreases as ζ increases for $n_{\mathrm{H}} \sim 10^4$ cm^{-3})
HNC	Sensitive to ζ (decreases as ζ increases for $n_{\mathrm{H}} \sim 10^4$ cm^{-3})

Here, χ and ζ measure the UV and the cosmic ray fluxes relative to their mean values in the interstellar medium of the Milky Way.

4.7 Further Reading

Cernicharo J., and Bachiller, R., eds. 2011. *The Molecular Universe.* IAU Symposium 280. Cambridge: Cambridge University Press.

Draine, Bruce T. 2011. *Physics of the Interstellar and Intergalactic Medium.* Princeton Series in Astrophysics. Princeton, NJ: Princeton University Press.

Hartquist, T. W., and Williams, D. A., eds. 1998. *The Molecular Astrophysics of Stars and Galaxies.* New York: Oxford University Press.

Hollenbach, D. J. and Tielens, A. G. G. M. 1997. Dense photodissociation regions. *ARAA* 35, 179–215.

Tielens, A. G. G. M. 2005. *The Physics and Chemistry of the Interstellar Medium.* Cambridge: Cambridge University Press.

5

Molecular Tracers in the Milky Way Galaxy

Studying the interstellar medium of the Milky Way Galaxy gives us the opportunity of identifying in detail the various components of the medium. The equivalent components in distant galaxies may be unresolved, but contribute to the overall emission. We show in Chapter 6 how to deal with emission from unresolved regions. In this chapter we consider the various distinct types of region in the Milky Way that can be explored through molecular line absorptions and emissions. We show that the chemistry in each of these molecular regions is dominated by one or more of the chemical drivers discussed in Chapter 3. The sensitivity of the chemistry to particular physical parameters, discussed in Chapter 4, may be an important concern in some cases. For most molecular regions, we identify a well-known example of each type, which is not necessarily typical but is one in which the consequences of the chemical driver are prominently displayed. We also list some molecular tracers useful in describing the physical conditions in these different situations. We emphasise in particular the tracers of density and temperature for Milky Way conditions. The aim of this chapter is to show how tracer molecules can reveal the nature, origin, and evolution of many types of region in the Milky Way. Tracers of conditions in galaxies external to the Milky Way are considered in Chapter 6.

5.1 Molecular Clouds

5.1.1 Diffuse Clouds

As we saw in Section 1.2, diffuse clouds (typically, $A_V \lesssim 1$ mag, n_{H} ~ 100 cm^{-3}, $T \sim 100$ K) gave the first indication that molecules might exist in interstellar space. Since then, diffuse clouds have been an important testbed for theories of interstellar chemistry. They also have an important role in the evolution of the interstellar medium. The molecular species that have been detected in diffuse clouds of the Milky Way are included in Table 1.1; these species are identified not only at UV and optical wavelengths but also in the infrared and millimetre regimes. In terms of the types of region that we described in Chapter 3, diffuse clouds may be described as photon-dominated regions (PDRs) of low density and low radiation intensity. The main chemical drivers in diffuse clouds are UV starlight and cosmic rays (see Section 3.3.2).

Specific example: A much studied line of sight through diffuse interstellar material is that towards ζ Oph. Visual extinction towards this star is 1 magnitude, and detections of very many of the diffuse cloud molecules listed in Table 1.1 have been made in the diffuse material on this line of sight. Conventional models are reasonably successful in accounting for the measured atomic and molecular abundances according to the chemistry described in Chapter 3 – with the notable exception of CH^+, H_3^+, and some other species – in terms of a static cloud with some denser substructure, subjected to fluxes of cosmic rays and starlight that are fairly close to their mean interstellar values. However, this picture cannot be complete because it seems clear from earlier discussions (Section 3.5) that a gas dynamic or other driver may be needed to produce CH^+ in the required amounts.

Many lines of sight through the diffuse medium also encounter transient microstructure – relatively tiny density enhancements. These were discovered through studying individual lines of sight and finding variations on timescales of around a decade or so in the intensity of atomic and ionic lines. These observations imply that structures typically of the size of the Solar System are present; they are denser than the background gas, overpressured, and therefore transient. Searches for molecules in these objects have detected CH^+, CH, CN, H_2CO, and possibly C_2. It is of interest to understand the implications for the dynamical processes that create such structures in the diffuse medium, and studying molecular absorption lines in the transient structures offers a useful approach. Models of diffuse clouds with transient structures suggest that CH^+ and HCO^+ may be formed in the transient structures, and that HS^+, CH_2^+, CH_3^+, H_2O^+, and H_3O^+ should also be present. Some of these species have been detected by the Herschel Space Observatory. Evidently, minor molecular ions may be good probes of the transient structures and therefore of the possible dynamical drivers that create them.

Some molecular tracers: We list here some important tracers of conditions in diffuse clouds. Because molecular emissions and absorptions in diffuse clouds occur in various parts of the electromagnetic spectrum, we also indicate the appropriate regime.

H_2 (UV): Rotational level populations can be determined from UV absorption spectra towards hot stars. The H_2/H ratio in diffuse clouds constrains the number density, temperature, and ambient UV radiation field. The high J level populations of H_2 are determined by the UV flux, by collisions with fast electrons, and by the formation of H_2 on dust grain surfaces. $H_2(J)$ observations may be able to constrain the parameters describing these processes.

HD (UV): Constrains the cosmic ray ionisation rate in clouds where HD does not contain all of the deuterium, and where the radiation field is known.

CO (UV) and **C_2** (IR): Rotational level populations of these molecules constrain the temperature. Chemical models to form these species constrain the number density and the radiation field.

CH^+ (optical): Detection of this species is an indication of nonthermal processes such as shocks or turbulence occurring in diffuse clouds along the line of sight, as temperatures $\sim$1000 K are required to form CH^+ (cf. Equation 3.33).

OH^+, H_2O^+, H_3O^+ (THz): These molecules are probes of clouds with small H_2/H values; they constrain the cosmic ray ionisation rate and number density, and may probe the transient microstructures.

NH, NH_2, NH_3 (UV and THz): The chemistry by which these species are formed in diffuse clouds is unclear, and they may be products of surface chemistry; observations of nitrogen hydrides towards W31C in absorption have revealed remarkable similarities among the profiles of these species, suggesting uniform abundances relative to hydrogen; they may constrain the radiation field.

HF (THz): This is the main reservoir of interstellar fluorine; HF is a useful probe of diffuse clouds with low H_2 fraction (the HF line saturates otherwise).

HCl (UV), **H_2Cl^+** (THz): These species are useful probes of the UV radiation field in diffuse clouds.

5.1.2 Molecular Clouds and Prestellar Cores

Molecular clouds are regions where most of the starlight is excluded by dust extinction ($A_V \gtrsim 5$ mag), and where – in the absence of extreme dynamical events or intense cosmic ray fluxes – most of the gas is molecular. Hydrogen is almost entirely in H_2, almost all the available carbon is in CO, and the excess oxygen appears in a range of other species. Quiescent molecular clouds are cooler and denser than diffuse clouds (typical values in molecular clouds are $T \sim 10$ K, $n_H \gtrsim 10^3$ cm^{-3}), and their chemistry is driven by cosmic rays rather than starlight. They almost certainly have high-density structures embedded in them; sufficiently dense clumps within them may become gravitationally unstable and initiate the early stages of star formation. Dust grains in molecular clouds become coated with icy mantles; if sufficient time elapses, then a significant proportion of all species other than H_2 and He will be contained in these ices.

Specific example: A well-studied molecular core is Barnard 68 (B68, Figure 3.1), a nearby ($\sim$125 pc) and yet opaque object ($A_V \sim$10 mags) that has been observed in several molecules including CO and its isotopologues, CS, HCO^+, and HCN. If one assumes a uniform gas-to-dust ratio, then the gas density profile for this core is very well constrained as a cloud in hydrostatic equilibrium confined by external pressure – i.e., a Bonnor–Ebert sphere – although recent observations of the line profiles of the CS (3–2) transition and the HCO^+ (4–3) transition indicate that B68 may not in fact be truly isothermal nor in hydrostatic pressure balance.

Some molecular tracers: Some important gas-phase tracers of density and temperature are listed here; all are detected in the submillimetre wavelength range:

CO and its isotopologues: In the absence of UV sources or of thermal excitation, hydrogen molecules provide no spectroscopic signature and CO molecules have long been used as proxy tracers of mass in molecular clouds. The CO(1–0) line is usually optically thick and therefore apparently a poor tracer; however, it has been found both from observational and theoretical studies that the CO(1–0) integrated line intensity ($\int T_a dv$, where T_a(K) is the antenna temperature and v (km s^{-1}) is the velocity; the integration is carried out over the linewidth) is roughly linearly related to the H_2 column density and that the approximate constant of proportionality in the solar neighbourhood is:

$$X = 2 \times 10^{20} \text{ cm}^{-2} (\text{K km s}^{-1})^{-1}. \tag{5.1}$$

However, X is rather sensitive to local physical parameters (see Figure 5.1). Similar relations can be derived using other rotational transitions of CO and of other species. Isotopologues of CO, that is, $^{12}C^{18}O$, $^{12}C^{17}O$, $^{13}C^{16}O$, $^{13}C^{18}O$, and $^{13}C^{17}O$, provide increasingly rare species that may generate optically thin (1–0) emission (in the Milky Way, typically, $^{12}C/^{13}C \sim$60, $^{16}O/^{17}O \sim$1500, and $^{16}O/^{18}O \sim$500). The CO(1–0) line is readily excited even at the low temperatures ($\sim$10 K) of molecular clouds, and is an effective tracer of interstellar gas at number densities $n_H \sim 10^3$ cm^{-3} but is not useful for tracing gas at much higher densities.

Tracers of structure within molecular clouds: As discussed in Chapter 3, dark clouds have a rich chemistry and many of the molecules detected are useful as tracers of gas at a higher density than that traced by CO(1–0) emission. High-resolution observations show that many dark clouds have a range of structures within them. These clumps may represent the very early stages of star formation, that is, prestellar cores. Different tracers appear to reveal different clumps.

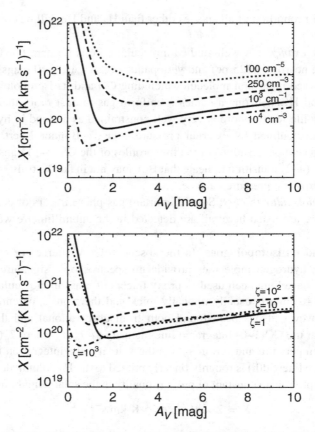

Figure 5.1. The X-factor versus A_V for (top) varying number density of hydrogen nuclei (in cm^{-3}); (bottom) varying cosmic ray ionisation rate (in 10^{-17} s^{-1}). The X-factor in both cases corresponds to the minimum of the curves, and is sensitive to these and other parameters. (Reproduced with permission from T. A. Bell, T. A., Roueff, E., Viti, S., Williams, D. A. 2006. *Monthly Notices of the Royal Astronomical Society*, 371, 1865.)

CO and its isotopologues trace the background gas in which structure is embedded; they give density and temperature estimates. They are also good tracers of depletion factors as it has been shown that, unlike H_2O ices for example, the CO ice abundance on different lines of sight may vary by factors up to 10 (see Figure 3.5).

$HCO^+(1–0)$, $N_2H^+(1–0)$: In quiescent clouds, HCO^+ and N_2H^+ emission provide constraints on the higher density structures in the cloud, and on the cosmic ray ionisation rate in cloud interiors. N_2H^+ rises as CO freezes out on to dust grains, so that the ratio of HCO^+ and N_2H^+ is sensitive to CO ice formation.

HCO^+ (3–2), CS(2–1): The line profiles of these transitions are good tracers of infall motion caused by gravitational collapse and possibly leading to star formation. However, the interpretation of molecular line profiles is complex due to the fact that infall signatures are similar to those caused by rotation and by oscillations. A good example of a molecular cloud where HCO^+ and CS maps reveal a complex pattern of red and blue asymmetric line profiles is the Bok globule Barnard 68. The CS (2–1) can also be used to constrain both the density and the available sulfur abundance – a poorly known quantity in dark clouds.

HCN, HNC, NH_3: These molecules are all good tracers of high-density structures; HCN is also a good tracer of infall, particularly for massive cores within molecular clouds, as revealed, for example, by Herschel observations of SgrB2(M).

Solid-state tracers: Dark clouds that are sufficiently opaque show solid-state features in the near infrared of molecular ices, predominantly H_2O, CO, CO_2, and tracers of other species (see Table 3.1). The ices indicate the presence of older, denser material (see Section 3.4.2).

Specific example of structure: L673 is a well-studied molecular cloud in emission from CO and its isotopologues. Fifteen small dense clumps within the cloud have been traced using interferometric arrays, by detecting emissions from NH_3, CS, HCO^+, and N_2H^+ (all known tracers of dense gas). The clumps have number densities $n_H \sim 10^4$ cm^{-3}, radii of a few percent of a parsec, and masses in the range $10^{-2} - 1$ $M_\odot$. The relation of these clumps to prestellar cores (see Section 5.3) is a question of topical interest that can be addressed through interstellar chemistry. The observations (see Figure 5.2) show that CS, HCO^+, and N_2H^+ appear to trace different clump distributions. This is now interpreted as a consequence of time-dependent chemistry occurring in collapsing cores at different phases of chemical and dynamical evolution, so that some molecular species (e.g., CS) peak in abundance before others (e.g., NH_3, N_2H^+, HCO^+).

5.1.3 Depletion in Dense Cores

In dense and cold cores, atoms and molecules in the gas phase freeze out on to the dust grains and are therefore lost, or depleted, from the gas. The extent to which this freeze-out (or depletion) occurs for a particular atom or molecule depends on a complicated chemistry that varies non-linearly with time and

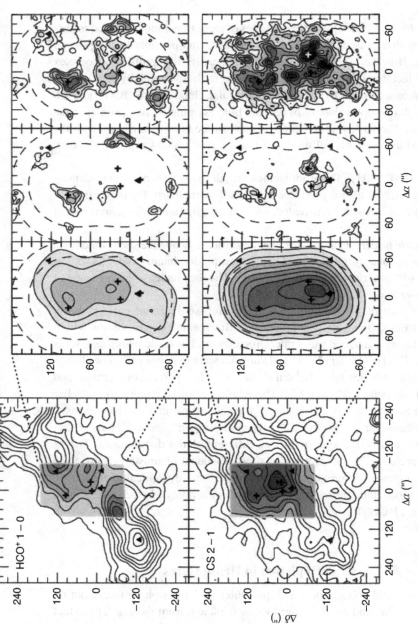

Figure 5.2. Contour maps in HCO$^+$(1–0)(top panels) and CS(2–1)(bottom panels) of the portion of the molecular cloud L673. The left-hand panels are 14-m single-dish (FCRAO) observations, and mid-left panels are the portion of these maps corrected for the primary beam of the BIMA interferometer maps that are shown in the mid-right panels. These (BIMA) interferometer observations miss a significant part of the molecular emission, and the right-hand panels show the combined FCRAO and BIMA maps. Evidently, this portion of the cloud is clumpy, but the HCO$^+$ and CS maps are not identical – a consequence of time-dependent chemistry. (Reproduced with permission from Morata, O., Girart, J.M., Estalella, R. 2005 *Astronomy & Astrophysics*, 435, 113. Copyright

68

physical environment (see Section 3.4.2). The strong dependence of depletion on the age and density of a core makes it a useful probe of core history.

It is difficult to quantify depletion observationally; CO gas emission is the most common molecule used to infer the fraction of species that is in the form of icy mantles. One does that by taking the ratio of the observed CO to the expected abundance at a particular density in steady state if freeze-out did not occur. This, however, implies a knowledge of the H_2 density as well as of the efficiency of nonthermal desorption mechanisms that would have returned the depleted CO in the gas phase regardless of the efficiency of freeze-out. In addition, the CO depletion factor is not necessarily equivalent to the molecular gas depletion factor, as different species freeze (and desorb) at different rates and have different (experimentally mostly unknown) sticking coefficients, as observationally shown by the Spitzer Space Observatory (see Section 5.3.1).

Nevertheless, it is worth noting that statistical comparisons of CO depletion among cores within a cloud, as well as comparisons among local molecular clouds, can help draw some general conclusions on the environment, and a recent study within the Gould Belt (see Figure 3.5) has found that within each cloud the highest levels of depletion are found in the longer lived regions, confirming, at least qualitatively, Equation 3.27; there is also a strong correlation between density and depletion, and starless cores are on average more depleted than protostellar cores. The fact that these conclusions are as expected from theory indicates that large comparative studies of CO can indeed be a useful tool to determine depletion in dense cores.

5.2 Star-Forming Regions and Their Outflows

Very young stars (sometimes called Young Stellar Objects, or YSOs) of both low and high masses tend to show outflow activity. In some cases the activity may extend to parsec-scale distances from the exciting star. The outflows may be highly collimated jets, or wide-angle winds. They are probably bipolar, although one lobe may be heavily obscured. Outflows of this type are seen in a variety of astronomical objects, including protoplanetary nebulae, X-ray binaries, symbiotic systems, and – on a galactic scale – active galactic nuclei. We focus here on outflows associated with young stars. Chemistry in these objects is largely dynamically driven, but in the case of massive stars their intense radiation fields also establish very extensive PDRs.

These outflows can be important sources of energy and momentum for the surrounding material, and both the flow properties and the ambient material can be studied through the interactions of the flows with their environments. These

Table 5.1. *Definition of Class 0 to Class III low-mass stars*

Class	n_H (cm^{-3})	T (K)	Age (years)	Disk
0	10^4–10^5	10	10^4	No
I	10^5–10^8	10–300	10^5	Yes (thick)
II	$>10^8$	>300	10^6–10^7	Yes (thick)
III	$>10^8$	>300	$>10^7$	Yes (very thin)

interactions arise through shocks between the outflows and the surrounding gas and through the entrainment of ambient material into the flows. The interactions create physical conditions that may generate a specific chemistry so that unique atomic and molecular tracers can be identified. Observations of the interactions through the atomic and molecular tracers are important because they help to define not only the nature of the outflows but also the properties of the gas in the environments of the young stars.

5.2.1 Outflows from Stars of Low Mass

The formation of a low-mass star (see Table 5.1) is believed to begin with the gravitational collapse of a rotating dense core in a molecular cloud (Class 0). Evidence suggests that the formation of a protostar at the centre of the core is accompanied by infall of material towards the equator and outflow from the poles (Class I). The T-Tauri phase (Class II) is reached when a disk has formed but infall and outflow are still occurring. Eventually the infall ceases, and the star approaches the main sequence with a remnant disk in which planets may form (Class III).

Collimated outflows, that is, jets, are observed to emerge from young low-mass stars, with jet velocities determined from proper motion studies to be in the range ~100–500 km s^{-1}. An example of a jet is shown in Figure 5.3. The very dense regions close to the star where jets originate can be probed by radio continuum centimetric emissions and by water masers. However, we shall concentrate here on the regions downstream from the jet launch region.

Emissions from [FeII], [SII], and from H$_2$ ro-vibrational and rotational lines are typically detected along jets and most especially at the heads of jets, where the most violent interaction occurs. Emission from the body of a jet occurs in regions that are less violently shocked than at the head, perhaps because periodic variations in the jet velocity enable faster material to overtake (and to shock) the slower material. In both cases the interaction creates a

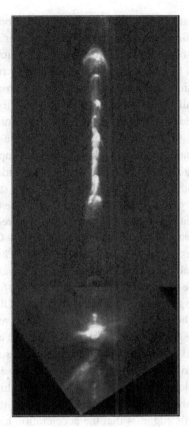

Figure 5.3. The Herbig–Haro object HH111 in Orion. The jet is about 4 pc in length and has internal structure, ending in a bow shock. The upper portion is a Hubble image in visible light; the lower portion is an infrared image showing a torus of dense gas around the star that generates the jet. (Courtesy of Bo Reipurth, CASA/U. Colorado, et al., HST, NASA.)

double-shock structure: a reverse shock slows the driving material and the forward shock accelerates the impacted material. The H_2 emission may be from molecules shocked and entrained in the jet, or from molecular hydrogen formed in the periodic double shock-system. The emission knots along a jet may also be traced in emission from $H\alpha$ and [FeII], as well as H_2 1–0 S(1) rovibrational emission. High-velocity CO is also found near the brightest emission regions.

Emission from the jets, and especially their heads (the so-called Herbig–Haro objects), may be intense enough to drive a photochemistry in nearby cold dense gas, creating characteristic tracer species. Detections of NH_3 and HCO^+ millimetre wave emission from clumps of dense gas near Herbig–Haro objects (see Figure 5.4) have been attributed to the release of icy mantles (including NH_3) from dust grains by photodesorption, followed by reactions between C^+ and H_2O (also from the mantles) to form HCO^+ at an abundance enhanced well above that normally found in molecular clouds. This chemistry produces a variety of other species that have also been detected. For example, along with HCO^+ detections, CH_3OH and H_2CO are observed to be strongly enhanced in clumps of dense gas near to Herbig–Haro objects, compared to normal dark cloud chemistry, while HCN and CN are found to be underabundant.

Classical bipolar outflows are found around young stars (later than Class 0). These outflows may be formed by jets as they erode the ambient medium through lateral shocks and entrainment, or possibly by precession of the star-jet system. The outflows can be traced in mature systems by CO and H_2 emissions where ambient gas has been entrained into the flow. The regions defined by CO with outflow velocity less than 30 km s^{-1} are generally poorly collimated and contain most of the outflow mass. The outflow lobes terminate in the shock (traced by H_2 IR emission and – optically – by Hα, [OIII], and [SII]) at the end of the flow). Higher velocity CO emission (with velocity >30 km s^{-1}) appears in the youngest flows (i.e., from Class 0 YSOs) and tends to be much more highly collimated than lower velocity emission; in some cases these flows appear to break up into 'molecular bullets', that is, dense and discrete clumps of high-velocity gas, detected in H_2 IR as well as CO (high *J*). The velocities may be much higher than the critical velocities for H_2 dissociation, in either C or J type shocks. This suggests that the acceleration has been gentle, rather than abrupt.

Other molecules are also useful tracers of bipolar outflows. In some sources, SiO emission is strong in both emission knots within the outflow and along the outflow walls. The high SiO abundance implied indicates that shock destruction of silicate grains may be occurring. The detected rotational lines of SiO require a hydrogen number density greater than about 10^6 cm^{-3}. The strongest SiO emission arises in the youngest sources. The outflow lobe walls are also traced by strong HCO^+ emission, where the HCO^+ is very much more abundant than in quiescent clouds. This overabundance appears to be a consequence of release of grain mantles from dust grains in this warm interface at the outer boundary of the flow, and the subsequent reaction of C^+ (from the outflow) with H_2O (from the mantles) to form HCO^+, a mechanism similar to that producing enhanced HCO^+ near HH objects.

In the youngest outflows (Class 0), infall on to the protostar may still be occurring. Both the outflow and the infall very close to the star may be traced by

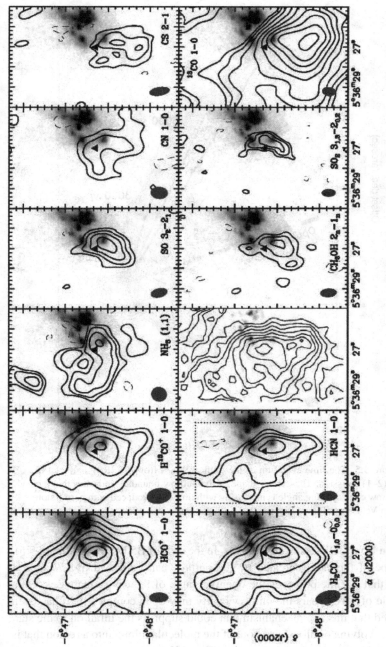

Figure 5.4. The association of emission from various molecules (as contours) with HH 2 is evident in these diagrams. The HH 2 emission is from SII and is shown in grayscale. The panel without a label is for dust emission at 0.85 mm. (Reproduced with permission from Girart, J. M., Viti, S., Estalella, R., and Williams, D. A. 2005. *Astronomy & Astrophysics*, 439, 601.) Copyright ESO.

73

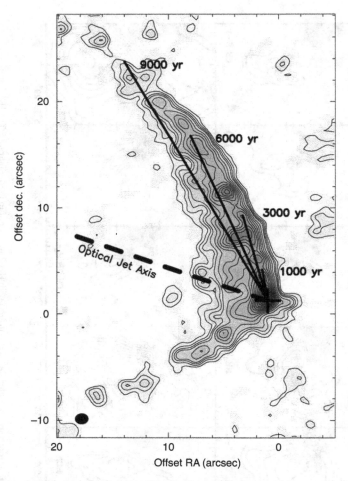

Figure 5.5. The time evolution of the outflow cones from B5 IRS1 is traced in CO(2–1) emission. The curved shape of the outflow boundary indicates that the outflow cone opening angle is increasing with time. (Reproduced with permission from Velusamy, T., and Langer, W. D. 1998. *Nature* 392, 685.)

emission from CO or its isotopologues. In the case of B5 IRS1 (see Figure 5.5), the shape of the cavity created by the outflow can be traced by CO, and it is clear that as time proceeds the opening angle of the cavity increases. On a timescale of $\sim 10^4$ years the outflow cavity may open completely. It has been suggested that this is a mechanism that could suppress the infall on to the star.

If the evolving outflow bursts out of the molecular cloud into a region that is largely atomic, then emission from the CO and H_2 tracers is no longer apparent.

In such cases, CO lobes cannot trace the full extent of the outflow, as defined by optical tracers. However, CO bullets in mature outflows may appear far beyond the extent of the CO outflows.

Tracers: H_2(IR), CO, and SiO (greatly enhanced) have already been mentioned. Many other species are substantially enhanced in abundance in shocks (as discussed in Section 3.5.1), including H_2O, OH, H_2S, SO, and SO_2, etc., and will trace activity in the outflows.

Specific example: One of the best examples of a chemically rich outflow is that originating in the Class 0 protostar L1157-mm (see Figure 5.6), at a distance of several hundred pc from the Sun. The outflow is associated with several bow shocks seen in emissions from H_2, CO, H_2CO, CH_3OH, and SiO. The brightest bow shock, L1157-B1, has been extensively studied and reveals a rich and clumpy structure as well as different molecular components at different excitation conditions coexisting within each clump. High spectral resolution observations with the Herschel Space Observatory provide excellent tracers of shock velocities, temperatures and pre-shock densities (via, e.g., NH_3 and H_2O), grain surface chemistry (via, e.g., H_2CO and CH_3OH), and grain sputtering efficiencies (via, e.g., HCl).

5.2.2 Outflows from Stars of High Mass

Regions of formation of massive stars also show dramatic evidence of the presence of outflows. Because massive stars are much brighter than low-mass stars, and mechanical luminosity tends to scale with stellar luminosity, the outflows around massive stars can be very powerful and extensive. Outflows in regions of massive star formation have all the characteristics that are observed in regions of low-mass star formation, that is, jets, winds, shocks, Herbig–Haro objects, bipolar structures, and bullets of very high velocity gas, and all of these may possibly be generated by more than one source in a stellar cluster. In general, the flows from the brightest sources (with luminosities $\sim 10^5$ solar luminosities ($L_\odot$)) are poorly collimated. Massive star-forming regions also show an additional feature to those found in regions of low-mass star formation: extensive PDRs are present, created by the powerful and hard radiation fields of the young massive stars. The chemical drivers in regions of formation of massive stars are therefore both dynamics and radiation. These regions also contain compact, dense clumps of gas and dust associated with the youngest stars, including so-called hot cores (described in Section 5.4.1). Regions of massive star formation are rare in the Milky Way and are generally deeply embedded in molecular clouds and very highly obscured.

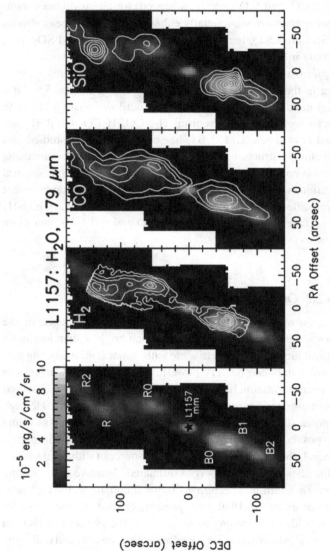

Figure 5.6. PACS map of the integrated H_2O 179 μm emission along the L1157 outflow from the Herschel satellite. The same map is shown in the other panels with overlays of other tracers, namely H_2 0–0 S(1) at 17 μm, CO 2–1, and SiO 3–2. The spatial resolutions of these images are ∼11″ for H_2 and CO and 18″ for SiO. (Reproduced with permission from Nisini, B., Benedettini, M., Codella, C., et al. 2010. *Astronomy & Astrophysics*, 518, 120.) Copyright ESO.

Massive stars-forming regions are some of the most complex regions to analyse.

Specific example: The Orion Kleinmann–Low complex (Orion KL) is, at about 450 pc from the Sun, the nearest and consequently best studied region of high-mass star formation (see Figure 5.7). The infrared luminosity of the main radiation source in Orion KL is $\sim 10^5$ $L_\odot$. The region has at least two outflows, one at relatively low velocity (~ 18 km s^{-1}) and one at high velocity (~ 100 km s^{-1}); their presence is confirmed through the detection of high-velocity wings in the emission lines from CO and from H_2S and other sulfur-bearing molecular species. These detections are clear evidence of shocks associated with the outflows. In the infrared, emissions from H_2O, OH, and CO give clear signatures of the expanding gas in the outflows, and [OI] and [CII] indicate the presence of a PDR. Strong SiO emission from both the low- and high-velocity outflows shows that the shocks are capable of disrupting silicate dust to provide gaseous SiO.

The outflows (the so-called spatially extensive 'plateau' component of Orion KL) are accompanied by many 'fingers' of emission from shock-excited H_2; proper motions show that the system of 'fingers' is exceptionally young – less than one thousand years old. Orion KL also contains the Orion Hot Core, a very dense ($n_H \sim 10^7$ cm^{-3}), warm ($T_K \sim 250$ K), and tiny region ($\lesssim 0.1$ pc) rich in complex organics and other species (see Section 5.4.1), and the Compact Ridge, a region somewhat less dense and much larger than the Hot Core. The whole complex is embedded in an extended ridge of quiescent, cooler (~ 60 K) and less dense ($\sim 10^5$ cm^{-3}) gas. The core of Orion KL also contains two compact HII regions.

Generally, bipolar lobes are expected to have a mixing layer at their boundaries, where turbulence entrains molecular gas into the lobe gas. The greatly enhanced abundance of HCO^+ in low-mass sources (described in Section 5.2.1) may be an indicator of that process, and a similar process may occur at the interface between outflows and molecular clouds in a region of formation of massive stars. Observations of Cep A East (see Figure 5.8) show the existence of an extended linear region with an unusual chemistry (present in H_2CS, OCS, CH_3OH, and HDO but absent in other conventional tracers – H_2S, SO_2, SO, and CS). This chemistry may reflect an interface zone in which grain mantles are evaporated into a warm, dense, and illuminated region.

Tracers of regions of massive star formation: These regions are locations in which the entire range of detected interstellar molecular species may be found. Thus, all known molecular species may be useful tracers. However, the 'plateau region' is well traced by simple species that trace shocks, such as H_2O, OH, and CO in the infrared.

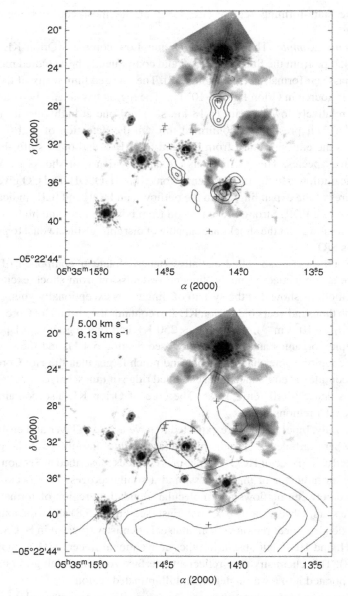

Figure 5.7. Molecular emission contours for (top) HCOOCH$_3$ and (bottom) HCOOH overlaid on the 2-μm gray scale that locates the main compact sources towards Orion KL. Note that methyl formate, (top), is rather concentrated while formic acid, (bottom), is extended. The plus signs denote the following sources (from E to W): Source I, Hot Core, SMA1, Source n, IRc7, Compact Ridge, IRc4, IRc5, IRc6, BN, and IRc3. (Reproduced by permission of the AAS from Widicus Weaver, S., and Friedel, D. N. 2012. *Astrophysical Journal Supplement*, 201, 16.)

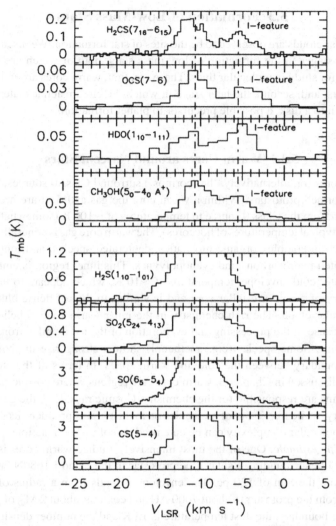

Figure 5.8. Molecular profiles towards Cep A East appear to fall into two categories, those that possess a feature, here called the I-feature, redshifted from the ambient velocity, and those that do not possess this feature. The species containing the I-feature appear in gas that is extended in space and is at least 0.1 pc in length. Chemical modelling indicates that the I-feature is neither a shock nor a hot core, but can be accounted for as an interface between an outflow and a molecular cloud. (Reproduced by permission from Codella, C., Viti, S., Williams, D. A., and Bachiller, R. 2006. *Astrophysical Journal*, 644, L41.)

5.3 Formation of Low-Mass Stars

Molecular clouds are observed to be the sites of star formation. We discuss here two key stages in the formation of low-mass stars, both of which are ideally suitable for study in molecular lines. These are, first, warm cores around young protostars, and, second, the later stage in which the circumstellar material has evolved into a disk, a possible precursor to planet formation.

5.3.1 Warm Cores around Nascent Stars

Warm cores (or, alternatively, 'hot corinos') surround Class 0 sources, that is, newly formed protostars. Within a warm core the gas and dust are heated by radiation from the Class 0 source to temperatures of ~ 100 K (somewhat cooler than the typical temperatures of hot cores). The warm core gas is generally found to be rich in complex organic molecules, deuterated species, and long-chain unsaturated hydrocarbons and cyanopolyynes. This inner region is embedded in an outer cold envelope, temperature of ~ 10 K, which appears to have the chemistry of a cold prestellar core and is similar to that of dense clumps in a molecular cloud. The rich chemistry in the inner warm zone is believed to arise because of the processing and evaporation of the ice mantles from warm dust grains. Some species (e.g., methanol) show a strong rise in abundance at the boundary between the cold and warm zones. In terms of the chemical drivers discussed in Chapter 3, warm cores are regions where cosmic rays and dust grains are responsible for the chemistry. Cosmic rays drive the gas-phase chemistry, while freeze-out of molecules on to dust grains produces ices within which molecular complexity can be enhanced by solid-state reactions.

Specific example: One of the most intensively studied warm cores is IRAS 16293-2422 surrounding a solar-type protostar in the cloud L1689N, at a distance from the Sun of 120 pc. The envelope extends from a radius of about 25 AU from the protostar to about 7000 AU and contains about 2 $M_\odot$ of gas. At the outer boundary, the dust temperature is 13 K and the number density is n_H $\sim 10^5$ cm^{-3}. The dust temperature rises to ~ 100 K at a radius of ~ 80 AU, where the number density is around 3×10^8 cm^{-3} and the ice mantles sublimate. The protostar is known to be a binary with separation between the components A and B of about 480 AU (see Table 5.2).

This source is chemically rich, and its spectrum has as many lines per GHz (about 20 in recent surveys) and as many identified species (more than 30) as a typical hot core (see Section 5.4.1). In addition, IRAS 16293-2422 is rich in rare isotopologues, especially in deuterated species. Emission from relatively simple molecules is significant, as is emission from complex organic

Table 5.2. Physical properties of IRAS 16293–2422A and
IRAS 16293–2422B

	IRAS 16293A	IRAS 16293B
n_H (cm^{-3})	4.9×10^7	7.9×10^7
Column density (cm^{-2})	5.9×10^{23}	7.1×10^{23}
M(M$_\odot$), of envelope	0.33	0.22

Data taken from Rao, R., Girart, J. M., Marrone, D. P., Lai, S.-P.,
Schnee, S. 2009. *Astrophysical Journal*, 707, 921.

molecules (COMs) as large as methyl formate (HCOOCH$_3$) and dimethyl ether
(CH$_3$OCH$_3$).

It appears that sources A and B of the binary may have different chemical compositions. Source B is brighter in emissions from complex organics,
whereas source A is brighter in emissions from simple N- and S-bearing species.
Dynamical considerations suggest that source A has a mass around 1 M$_\odot$, while
source B may be more than one order of magnitude smaller. Thus, warm core
chemistry may be linked directly to protostellar mass. Suitable molecular probes
of warm cores may even be able to constrain the outcome of the star formation
process.

While unbiased spectral surveys of warm cores have been exceptionally
useful in revealing the wealth of the chemistry, they also reveal that individual
species can be used to investigate particular characteristics.

Molecular tracers: CH$_3$OH, H$_2$CO, H$_2$O, and SiO show a strong rise at the
boundary of the warm core and can be used to constrain chemical models of
the warm phase.

HDO, D$_2$CO, and many other deuterated species constrain the chemistry in
the cold prestellar core phase. Complex organic molecules (COMs) for example, CH$_3$OH, HCOOCH$_3$, CH$_3$CN, and (CH$_2$OH)$_2$, constrain the chemistry
during the warming-up phase; they may be preferentially associated with less
massive Class 0 sources.

Carbon chains and cyanopolyynes constrain gas-phase chemistry in the
warm core.

The detection of COMs in warm cores is of particular interest because
there are no efficient gas-phase chemical pathways to form them under the
conditions in which they are observed. It appears that they are formed in solid-
state chemical processing of ices that were deposited on dust grains during the
collapse phase of star formation and in gas-phase reactions between products

of that processing after evaporation. Therefore, detections of large molecules
in warm cores (and in hot cores; see Section 5.4.1) help to probe the solid-state
chemistry and the physical conditions in which the ices were deposited and
chemically processed.

In recent years, especially with data from the Spitzer Space Observatory,
solid-state molecular features have been observed in many cores and protostars,
enough in fact to be able to draw some conclusions on the nature of the ices
in these objects (see Table 3.1). Water is of course the most abundant ice
component, with the abundances of CO and CO_2 being $\sim$30% that of water.
CH_4 and NH_3, on the other hand, are quite low in abundance (by at least a
factor of 5 less than CO). Whereas CH_4, NH_3, and H_2O vary little across a
large sample of low-mass protostars, CO, CO_2, and CH_3OH vary by factors of
2–3 or more. Finally, there is some evidence of more complex ices, as predicted
by chemical models.

5.3.2 Protoplanetary Disks

Protoplanetary disks are an important stage of development between warm
cores and the formation of planetesimals and planets. The disks are observed
around low-mass (T Tauri) and intermediate-mass (Herbig Ae) stars. The chemistry of the disks helps to constrain the molecules that may be incorporated
in protoplanets. Disks traced in CO emission typically have radii less than
$\sim$10³ AU, subtending less than 1 arcsec at 1 kpc distance so that array telescopes operating in millimetre and submillimetre wavebands are normally
required to resolve disk structure.

A protoplanetary disk may be strongly illuminated by UV and (usually vari-
able) by X-ray emission from the central protostar, and (because star formation
often occurs in clusters) possibly by radiation from similar objects nearby. The
disk may also be affected by UV line emission generated in the dissipative
accretion process, especially in the case of more massive (Herbig Ae) stars.
Material in the disk close to the source (at a radius of a few AU) is warm
and radiates strongly in the infrared, while material further from the source is
cooler and emits at millimetre wavelengths. A gas-rich disk (see Figure 5.9) is
normally flared and will have a PDR or XDR (or both) created on its outer faces
by UV and X-radiation from the central source, or from nearby sources. Warm
molecular layers lie beneath these outer PDR/XDR zones, and a midplane layer
will, if mixing is slow, be cold enough at large radii for extensive freeze-out of
gas phase species to occur. This, and the small size of the objects, means that
molecular column densities in disks are generally low compared to those in
warm cores. In terms of the chemical drivers discussed in Chapter 3, the disks

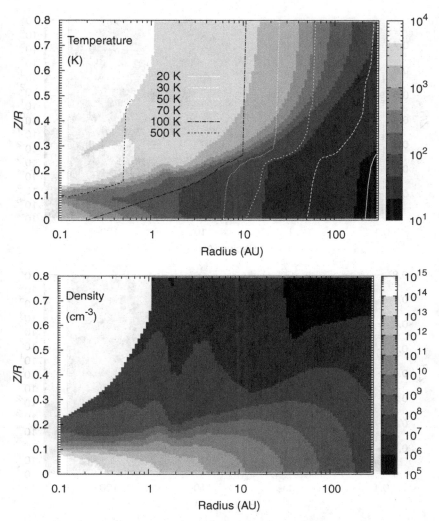

Figure 5.9. The panels show the temperature, density, and UV and X-ray fluxes in a circumstellar disk around a model T Tauri star, computed as a function of disk radius, R, and height, Z, above the disk. There is an extremely wide range in all parameters. The above disk material is strongly irradiated by UV and X-rays, and is both a PDR and an XDR, while material close the plane is very cold and dense and almost radiation-free. On the temperature panel, the gray scale represents the gas temperature, while the contours represent the dust temperature. (Reproduced with permission, from Walsh, C., Nomura, H., Millar, T. J., Aikawa, Y., 2012, *Astrophysical Journal*, 747, 114.)

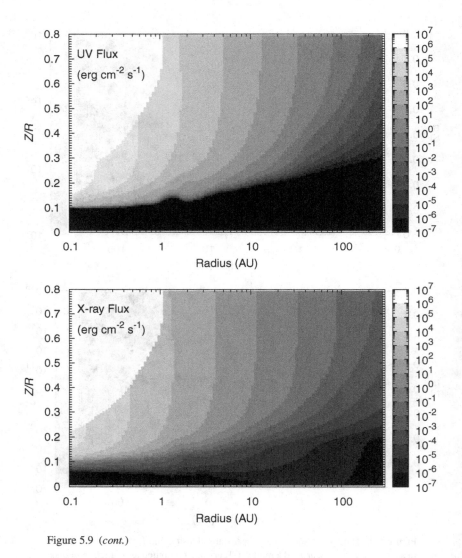

Figure 5.9 (*cont.*)

are complex regions where cosmic ray–induced reactions, PDRs, XDRs, and ice processing on dust grains all contribute to the chemistry.

Specific example: There is a significant range of chemical properties across observed sources, and these variations provide clues to the chemical drivers and source of excitation. A 'typical' disk cannot, therefore, be identified. Molecular detections in a number of disks are shown in Table 5.3. A relatively rich chemistry is found to be present in IM Lup (not listed in the table), a source

Table 5.3. Abundances of some gaseous molecules detected in disks around T Tauri and Herbig Ae stars

Molecule	LkCa15	TW Hya	HD 163296	MWC 480	DM Tau
CO	3.4×10^{-7}	5.7×10^{-8}	3.1×10^{-7}	6.9×10^{-7}	9.6×10^{-6}
HCO^+	5.6×10^{-12}	2.2×10^{-11}	7.8×10^{-12}	1.0×10^{-10}	7.4×10^{-10}
DCO^+	$<2.31 \times 10^{-12}$	7.8×10^{-13}	–	–	–
CN	2.4×10^{-10}	1.2×10^{-10}	1.3×10^{-10}	1.4×10^{-10}	9.0×10^{-9}
HCN	3.1×10^{-11}	1.6×10^{-11}	$<9.1 \times 10^{-12}$	$<1.1 \times 10^{-11}$	4.9×10^{-10}
HNC	–	$<2.6 \times 10^{-12}$	–	–	1.5×10^{-10}
CS	$<8.5 \times 10^{-11}$	–	–	–	2.4×10^{-10}
H_2CO	4.1×10^{-11}	$<1.4 \times 10^{-12}$	$<1.0 \times 10^{-11}$	$<1.4 \times 10^{-11}$	2.4×10^{-10}

Data taken from Thi, W.-F., van Zadelhoff, G.-J., van Dishoeck, E. F. 2004. *Astronomy and Astrophysics*, 425, 955.

with a central object of type M0 that has near-Solar luminosity, and a large complete disk with mass of about 0.1 $M_\odot$. Detected molecules in the disk of IM Lup include CO, HCO^+, DCO^+, N_2H^+, H_2CO, HCN, and CN at submillimetre wavelengths, and CO_2, H_2O, OH, HCN, and C_2H_2 at infrared wavelengths. In general, disks around T Tauri stars are more chemically rich than those around Herbig Ae stars.

Useful tracer molecules: CO measures the physical characteristics of disks and their orientation; the HCN:CN ratio is used as a measure of the local radiation field. Measurements across a range of sources can help to determine the contribution of the radiation generated during the accretion process to the total stellar radiation. The HCN:CN ratio is also influenced by the effects of dust settling and grain growth that allow more radiation to penetrate the disk. The ratio of HCO^+ to DCO^+ probes the deuterium fractionation efficiency, a strongly temperature-sensitive effect. The ratio of N_2H^+ to HCO^+ traces the freeze-out of CO because N_2H^+ is destroyed in reactions with CO. H_2CO indicates the presence of an extensive cold chemistry in a disk.

5.4 Formation of High-Mass Stars

As described in Section 5.2.2, regions where massive stars are formed contain a remarkable variety of molecular sources, including the cool background molecular cloud gas with embedded dense molecular cores, the jets and winds from young stars driving shocks and molecular outflows, the interaction of stellar radiation with the cloud gas to generate a PDR, and the very dense circumstellar matter. In recent years, it has been suggested that the earliest stage of high-mass star formation occurs within the so-called infrared dark clouds (IRDCs). These clouds are regions of high extinction viewed against the bright diffuse mid-infrared galactic background: the molecular material in these regions is cold (< 25 K), dense ($n_H > 10^5$ cm^{-3}), and has high column densities ($N_H \sim 10^{23}$–10^{25} cm^{-2}).

We highlight in this section a highly characteristic signature of massive star formation: the molecular-rich dense hot cores.

5.4.1 Hot Cores Near Massive Protostars

Hot cores are small ($\lesssim 0.1$ pc), dense ($n_H \gtrsim 10^7$ cm^{-3}), and warm ($T_K \sim 200$ K) objects found in the vicinity of newly formed massive stars. They are exceptionally rich sources of all types of molecules, including many isotopologues, and are particularly remarkable for the wealth of COMs detected in them. Hot cores are fragments of very dense gas associated with the formation of a massive protostar during the collapse of a molecular cloud. These fragments are infalling material very close to the protostar but not incorporated into it. The material is, however, strongly irradiated and warmed by the nearby protostar. Molecules are therefore useful probes of hot cores themselves and – with the aid of models – of the earlier cold phase of the collapse from a less dense state. The chemical drivers in hot cores (cf. Chapter 3) are therefore cosmic rays (inducing gas-phase chemistry) and dust grain processes that lead to the formation of ices and their subsequent processing.

The abundances of species detected in hot cores can be surprisingly large; for example, the fractional abundances of OCS and CS in the Orion Hot Core are both near $\sim 10^{-7}$. Different molecular species appear to probe different parts of the hot core. For example, methyl cyanide in the hot core G31.41+0.31 is relatively widely distributed, whereas glycol aldehyde is much more concentrated to the centre of the core, in contrast to the molecular distribution in the Galactic Centre (see Section 5.6).

Specific example: G29.96-0.02, a bright hot core at a distance of ~ 6 kpc, is a strong molecular line emitter (see Figure 5.10 and also Figure 1.3). In this source, many types of molecular species have been detected in large abundance, including NH$_3$, CH$_3$CN, HNCO, HCOOCH$_3$, CH$_3$OH, H$_2$CO, H$_2$S, and others. CH$_3$OH, with its forest of lines, allows a determination of the temperature gradient in this hot core (≥ 300 K), while sulfur-bearing molecular line observations give support to the view that time-dependent chemical desorption is occurring from the grains. Besides being a prototypical hot core, G29.96-0.02 is also interesting because it is one of a handful of hot cores in which clear evidence of

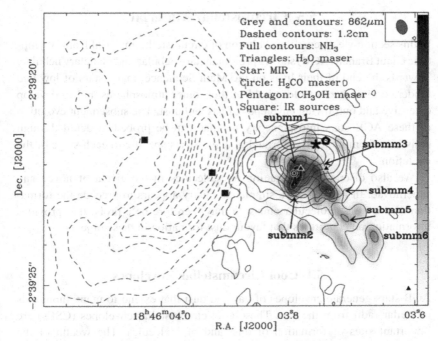

Figure 5.10. The hot core and ultra-compact HII region G29.96-0.02. The grayscale with contours shows the submillimetre continuum emission from warm dust. The dashed contours are of centimetre continuum emission from the HII region. The full contours are of NH₃ emission. Other sources are noted on the diagram. (Reproduced with permission from Beuther, H., Zhang, Q., Bergin, E. A., Sridharan, T. K., Hunter, T. R., and Leurini, S. 2007. *Astronomy & Astrophysics*, 468, 1045.) Copyright ESO.

rotation and associated bipolar outflows has been found via the interpretation of high J molecular transitions (such as CH_3CN $J > 6$ observations).

Tracer molecules: All COMs provide observational information on the gas dynamics within the core and on the spatial distribution of molecular species within the core. They can therefore constrain models for molecule formation in hot cores. The high dipole moment methyl cyanide molecule is an excellent tracer of the kinematical structure, while the large number of methanol transitions allows a determination of the temperature structure. Finally, sulfur-bearing species, with their transient chemistry, may be useful evolutionary tools.

Observations of ices towards high-mass protostars reveal lower levels of CO and CO_2 ices than those in the low-mass counterparts (see Table 3.1), no doubt due to the stronger heating of the dust towards high-mass stars.

5.5 Circumstellar Material

In this section we discuss the evolution of circumstellar material from Asymptotic Giant Branch (AGB) stars to protoplanetary nebulae and planetary nebulae. Towards the end of their lives on the Main Sequence, many stars of low and moderate mass begin to lose hold of their outer atmospheres and to develop extensive and outflowing envelopes of gas and dust. The subsequent evolution of these AGB stars into planetary nebulae can be probed in detail through emissions from molecules formed in chemistries specific to each stage of the evolution.

We also discuss the formation of molecules in the ejecta of novae and supernovae. In both cases, a simple chemistry appears to precede the formation of dust, and molecular emission may therefore be a probe that provides information on conditions that favour dust formation in these objects.

5.5.1 Cool Circumstellar Envelopes

AGB stars generate envelopes of gas and dust that extend to many thousands of stellar radii from the star. These cool circumstellar envelopes (CSEs) are important sites of formation of dust and of molecules. The wealth of the chemistry in these objects is surprising, and includes simple inorganic species, organics, radicals, rings, and chains (see Table 1.1). These envelopes are drifting away from the star with typical radial velocities around 10–20 km s^{-1}. Each flow is fed at the base with material from the cool stellar atmosphere, containing (for the case of stars with C/O > 1) fairly simple species such as H_2, CO, C_2, CN, HCN, and C_2H_2 generated in local thermodynamic equilibrium (LTE) in the stellar atmosphere. The drift is caused by pulsation-driven shocks occurring at a few stellar radii, and the outflow may be clumpy. From this radius outwards to about a hundred stellar radii is the region of dust formation. Beyond this region is a molecular-rich zone, possibly extending to some tens of thousands of stellar radii.

These systems are therefore rather complex ones in which a dusty gas, rich in simple molecular species, is moving into a region dominated by the effects of the interstellar radiation field. Therefore, it is essentially a photon-dominated region (a PDR; see Section 3.2.1) in a molecular gas that is moving into a radiation field that grows stronger as the gas (and dust shielding) becomes more dilute; eventually, the radiation field attains its external interstellar strength, destroys the molecules, and returns the gas to an atomic/ionic state. The dust may be largely unaffected by the radiation field and eventually mixes with the interstellar gas. The photochemistry of the 'parent' species from the stellar atmosphere (CO, CN, C_2H_2, etc.) gives products that take part in a chemistry, in 'the photochemical zone' out to some thousands of stellar radii, to create 'daughter' species in the expanding gas.

Although these envelopes appear more complicated than some other locations of cosmic chemistry such as static interstellar clouds, in fact these are systems in which precise observations and modelling of molecular astrophysics can be performed. These studies yield important information on the structure and dynamics of CSEs, on stellar evolution, and on the injection of material (especially dust) into the interstellar medium. The precision of the models arises because – in comparison with models of interstellar cloud chemistry – there is a well-defined and single source of material (the cool star) and the geometry is well defined and simple, so the physical parameters are mostly well determined. The chemical timescale is determined by the dynamics, the density, and the external radiation field, and is (typically) on the order of ten thousand years. Figure 5.11 illustrates schematically the variety of processes occurring in a cool carbon-rich stellar envelope.

Specific example: IRC+10216 is a well-studied and nearby ($\sim$100 pc) carbon-rich AGB star. Its envelope has a particularly rich chemistry; more than 70 molecular species have been identified in it. The proximity of IRC+10216 allows a study of the internal structure of the CSE, including the formation of H_2O, SiO, and SiS. The wider structure shows evidence of episodic ejection events rather than a uniformly steady flow.

Useful tracers: H_2O, OH, and H_2CO can be present even in carbon-rich objects, if the envelope is clumpy, giving information on the low-velocity shocks. Carbon chains such as HC_3N may also be enhanced through clumpiness, so that envelope gas is more shielded from the destroying effects of the radiation field. The clumpiness in the envelope indicates possible episodic ejection.

The anions C_4H^-, C_6H^-, C_8H^-, CN^-, C_3N^-, and C_5N^- have been detected in IRC+10216 and several of these anions have been found in other CSEs (though not elsewhere in the interstellar medium; see Table 1.1). Their simple chemistry gives an accurate test of the chemical network used in an expansion model and the evolution of the chemistry in the envelope.

5.5.2 Planetary Nebulae

The high-mass loss rate ($\sim 10^{-4}$ $M_\odot$ y^{-1} for IRC+10216) in the AGB phase obviously cannot last very long. When it ceases, the object is termed a Proto-Planetary Nebula (a PPN), an object totally enveloped in and optically obscured by its own circumstellar dust and detectable only in the far infrared. Once

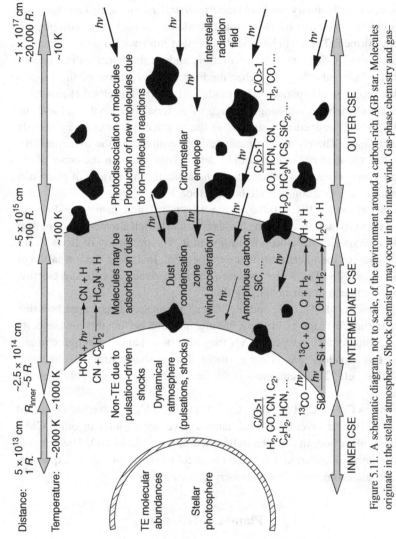

Figure 5.11. A schematic diagram, not to scale, of the environment around a carbon-rich AGB star. Molecules originate in the stellar atmosphere. Shock chemistry may occur in the inner wind. Gas-phase chemistry and gas–grain interactions occur farther out. The penetration of interstellar UV into the envelope drives a photochemistry in the outer regions. (Reproduced with permission from Decin, L., Justtanont, K., De Beck, E., et al. 2010. *Nature*, 467, 64.)

the flow has ceased, the central star contracts and eventually becomes so hot ($\sim$30 000 K) that its radiation is hard enough to photoionise the surrounding nebula. Simultaneously, the star generates a very fast ($\sim$2000 km s^{-1}) low-density wind that compresses and shapes the surrounding nebula gas into a Planetary Nebula (a PN), frequently into a toroidal structure with bipolar outflows. The timescales associated with these three phases, AGB envelopes, protoplanetary nebulae (PPNe), and planetary nebulae (PNe), are roughly 10^4 yrs, 10^3 yrs, and 10^4 yrs, respectively. Observations of molecules present in these objects therefore give an opportunity to follow evolutionary events on these astronomically very short timescales. PPNe and PNe are thus rather complex objects; they include regions affected by strong UV from the central star (i.e., PDRs) and by the fast stellar winds and the shocks they induce (dynamically driven regions), and outer regions that remain similar to AGB cool envelopes.

The evolution of an AGB into a PPN and then a PN is accompanied by an evolution of the chemistry of gas-phase molecules. The molecular component may be dominant during a large part of the evolution of the nebula and hence plays an important role in their mass distribution and shaping. A similar evolution occurs in the chemical structure of amorphous carbon dust grains. This evolution can be followed in the near-IR 3-μm band, which shows the conversion of aliphatic (H-rich) structures in AGBs being converted to aromatic (H-poor) structures in PPN and PN, by the influence of the radiation of the central star.

Specific example: CRL 618 is a PPN that has evolved very recently from the AGB stage (see Figure 5.12). It has a B0 central star, a dense torus, an envelope of mass $\sim$1 M$_\odot$ expanding at about 20 km s^{-1}, and high-velocity clumpy gas expanding at $\sim$200 km s^{-1}. This PPN is rich in molecules (see Table 5.4), many of which are detectable in ro-vibrational absorption in the infrared. Some of the molecules are typical of the AGB stage, whereas others are the products of chemistry initiated by the stellar UV and shocks. Carbon chains, such as C_4H_2 and C_6H_2, are abundant and their formation is driven by C_2H addition reactions, this radical being create by UV photodissociation of C_2H_2 in the inner regions of the PPN. High J lines of CO trace the slow expansion of the central densest parts created by the final outflow from the (previously AGB-type) star. Methyl-substituted polyynes, such as CH_3C_2H and CH_3C_4H, are present in CRL 618, but not found in the prototypical AGB, IRC+10216. Ring structures, such as benzene, are also found in CRL 618. Of particular interest are some species, such as HCO^+, CO^+, HCN, and CN, which are found to be particularly abundant compared to their values in interstellar clouds. It is unclear how these species can survive the strong radiation field to which they are subjected during

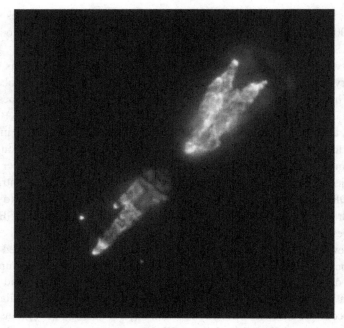

Figure 5.12. CRL 618 is an aspherical protoplanetary nebula. The optical and UV emission from the central star is obscured by dust, and the star has moved off the Main Sequence. The ejection of gas and dust at speeds up to 200 km s^{-1} creates the nebula and began about 200 years ago. (Courtesy of ESA/Hubble & NASA.)

the expansion; one possibility is that they are well protected in dense clumps (such as the famous cometary knots in the Helix nebula).

Useful tracers: Various hydrocarbons and ions, such as HCO$^+$ and CO$^+$, are useful tracers of the AGB to PPN transition. The appearance, or enhancement, of HCO$^+$ may correspond to different evolutionary stages, depending on what causes its high abundance: at the early stages this ion can be formed via reactions of H$_3^+$ (short-lived) with CO, whereas at later evolutionary stages reactions of CO$^+$ with H$_2$ may be responsible. The varying ratios of HCO$^+$ and CO$^+$ in relatively large samples of PPNe and PNe could potentially be used as an evolutionary diagnostic tool. Lines of H$_2$ may instead be good tracers of density as they indicate emission from knots of very dense gas ($n_H = 10^5$–10^6 cm^{-3}) embedded within ionised gas; the outer layers of the knots are PDRs excited by the radiation from the central star. Lines from the fullerenes C$_{60}$ and C$_{70}$ have been detected in a young PN, suggesting that hydrocarbon chemistry is very rapid.

Table 5.4. Column densities of a sample
of molecules observed in CRL618

Molecule	Column density (cm^{-2})
CN	6.5×10^{14}
CS	1.9×10^{13}
HNC	4.2×10^{13}
HC_3N	1.2×10^{14}
SiO	3.9×10^{12}
HCO^+	1.6×10^{14}
CO^+	3.0×10^{12}

Data taken from Bachiller, R., Forveille, T., Huggins, P. J., and Cox, P. 1997. *Astronomy & Astrophysics*, 324, 1123 and Bell, T. A., Whyatt, W., Viti, S., and Redman, M. P., 2007, *Monthly Notices of the Royal Astronomical Society*, 382, 1139.

5.5.3 Ejecta of Novae

H_2, CO, CN, SiO, SiO_2, and SiC have been detected in the ejecta of novae through the ro-vibrational emission of these species in the near-infrared. The molecules are found to be present at early times ($\sim$10 days post-outburst) in the evolution of the nova, and – in particular – before the formation of the optically thick dust shells that are seen in about a third of all novae, at a time, typically, of about a month after outburst. Thus, molecule formation seems to be a precursor to dust formation in novae. The physical conditions in which this chemistry occurs are extreme compared to interstellar and other circumstellar situations, including very high but rapidly declining number densities, very high but declining initial temperatures, and intense and hardening radiation fields as the central star evolves. Molecule and dust formation appear possible only in a transient zone where cooling and total line blanketing by Fe allow chemistry to operate in the expanding shell of ejecta. Thus, although intense radiation fields are certainly present in the molecular zone, chemistry occurs in sufficiently dense regions in which the radiation does not dominate. The main driver of chemistry in novae is, therefore, the initial explosion that shocks, heats, ionises, and expels the gas. When recombination of ions and electrons has proceeded almost to completion, conditions permit the formation of molecules. We conclude that the driver of chemistry in novae is therefore dynamical. The chemistry involves neutral-neutral and ion–molecule reactions, using the residual ionisation remaining during the period of recombination, followed by dust formation. The types of dust detected through IR spectroscopy of novae include amorphous carbons, hydrocarbons, silicates, silicon carbide, and alumina.

Specific example: QV Vul(1984) showed molecules as early as 20 days days after outburst. The first dusts to appear were amorphous carbons, silicon carbide, and hydrocarbons at $\sim$50 days, whereas O-rich silicates appeared later ($\sim$100 days).

Molecular tracers: H_2, – CO, CN, SiO, SiO_2, SiC, and C_2 – may also be abundant.

5.5.4 Ejecta of Supernovae

The environments of supernovae may seem unlikely locations for the formation of molecules. Yet detections of CO and SiO have been made in the ejecta of several Type II supernovae at 100 days or more after the outburst. Dust formation may follow somewhat later (perhaps around a year after outburst). The molecular fraction of the ejecta may be about 30% by mass.

The precursor stars of Type II supernovae are supergiant stars of masses of $10\,M_\odot$ or more. At the end of the life of the star, the stellar core contains the products of nuclear synthesis in a number of distinct layers with unburnt hydrogen on the outside. Helium is in the first shell inside the hydrogen layer, then carbon and oxygen, and so on, with iron in the centre. If there is no mixing of layers during the explosion, then these shells remain distinct and the chemistry in the ejecta therefore involves only those elements belonging to each particular shell. The ejecta are subjected to gamma-ray fluxes from the radioactive decay of ^{56}Ni and ^{56}Co, and to UV and other fields. The ejecta are initially very dense ($\sim 10^{12}$ cm^{-3}) and hot ($\sim 10^4$ K), and 3-body chemistry will occur. In terms of chemical drivers (Chapter 3), these regions are very high-density PDRs with an unconventional radiation field. The chemistry is also strikingly unconventional, as hydrogen (normally the dominant element in all the chemistry in the Universe) is absent, unless mixing of shells occurs. Chemical routes are, therefore, highly restricted compared to those in H-rich environments. The most important shell for the resulting chemistry is that containing carbon and oxygen.

Specific example: Molecules CO and SiO were detected in SN1987A by their fundamental emissions at 416 μm and 8μm, respectively. The CO emission was detected from 157 to 532 days after outburst. The SiO emission was detected from 160 to 519 days, and the onset of dust formation was detected at about 530 days, suggesting that silicon was incorporated into dust grains at that time.

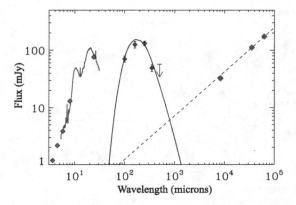

Figure 5.13. The infrared spectral energy distribution of SN 1987A. Herschel detections were made between 100 and 300 μm, and an upper limit was made at 500 μm. Emission from warm dust and the radio continuum are also shown. (M. Matsuura, private communication.)

The detected dust mass was initially $\sim 10^{-4}$ $M_\odot$, but recent observations by the Herschel Space Observatory show that by 2011 this had increased to 0.4–0.7 $M_\odot$. The dust composition includes both amorphous carbon and silicates and incorporates most of the refractory material available. Figure 5.13 shows the infrared spectral distribution of SN 1987A.

Tracer molecules: SiO and CO constrain the expansion model. Detection of other species might indicate that mixing between layers occurred during the explosion.

5.6 The Galactic Centre

The Central Molecular Zone (CMZ) is a half-kiloparsec region at the centre of the Milky Way Galaxy (see Figure 5.14) in which the physical conditions of the gas are strikingly different from those in the galactic disk. Gas clouds of average number density $n_H \sim 10^4$ cm^{-3} and of size ~ 20 pc exist with kinetic temperatures ~ 200 K. Dust grain temperatures within these clouds are $\lesssim 30$ K. These clouds are rich in molecules (see Figure 5.14) and are found to be locations for the most complex organic molecules so far identified in interstellar space; these COMs have a high abundance relative to hydrogen, typically 10^{-9}–10^{-8}; the relative abundance of methanol is even higher, 10^{-7}–10^{-6}. It is of interest to characterise better the population of clouds and to identify the factors that make this region so different from other molecular regions in the Galaxy. Molecular emissions from the CMZ are important contributors in this work.

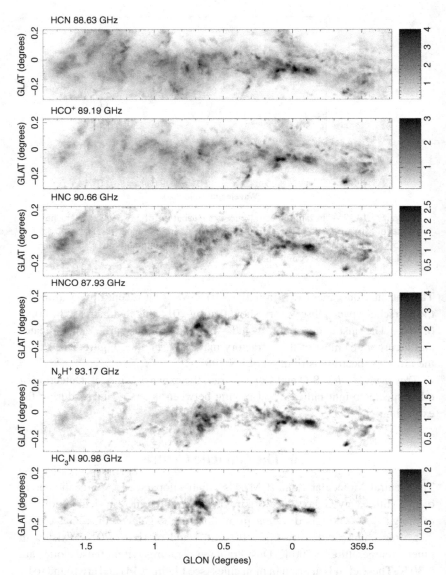

Figure 5.14. Maps of the CMZ in various molecular species. The grayscale represents the antenna temperature in Kelvin. GLAT and GLON are the galactic latitude and longitude, respectively. (Reproduced with permission from Jones, P. A., Burton, M. G., Cunningham, M. R., Requena-Torres, M. A., Menten, K. M., Schilke, P., Belloche, A., Leurini, S., Martin-Pintado, J, Ott, J., and Walsh, A. J. 2012. *Monthly Notices of the Royal Astronomical Society*, 419, 2961.)

The CMZ is a very active zone in the Milky Way Galaxy, and its chemistry is complex and may involve all four of the main drivers discussed in Chapter 3. This is certainly a region of enhanced cosmic ray fluxes and intense radiation fields in which violent dynamical processes are also occurring. The presence of abundant COMs that cannot be formed in gas-phase reactions suggests that dust grain processes may play a role. Because of the complexity of the CMZ, this section provides a link to the discussions in the next chapter.

Specific example: One galactic centre cloud that is particularly rich in molecular species is MC G-0.11-0.08. Molecules detected in this cloud, with relative abundances in the range stated earlier, include the following: CH_3OH (methanol), C_2H_5OH (ethanol), $(CH_3)_2O$ (dimethyl ether), $HCOOCH_3$ (methyl formate), $HCOOH$ (formic acid), CH_3COOH (acetic acid), H_2CO (formaldehyde), CH_2CHO (propynal), CH_2CHCHO (propenal), CH_3CH_2CHO (proprionaldehyde), CH_2OHCHO (glycolaldehyde), $HOCH_2CH_2OH$ (ethylene glycol), c-C_2H_4O (ethylene oxide), CH_3CHO (acetaldehyde), and H_2CCO (ketene).

These molecules are generally regarded as signatures of hot cores (see Section 5.4.1), and the abundances of CMZ molecules relative to methanol are strikingly similar to those in hot cores. However, whereas hot core molecules arise mainly from solid-state chemistry followed by the evaporation of ices and from a compact region, in the CMZ the molecular emission tends to be extended. In fact, the physical conditions in hot cores (number density n_H $\sim 10^7$ cm^{-3}, size $\lesssim 0.1$ pc) and in the CMZ ($n_H \sim 10^4$ cm^{-3}, size 20 pc) are very different. The chemistry that produces molecules in small-scale hot cores may not be the same as that in large-scale CMZ clouds. Grain temperatures in the CMZ are low enough that evaporation of mantles may not occur, so the erosion of ice mantles in frequent low-velocity (~ 20 km s^{-1}) shocks has been invoked.

A more typical hot core is also found close to the Galactic Centre; this is probably the richest in molecular species known in the Galaxy, and has been dubbed the Large Molecule Heimat (Sgr B2(N-LMH)). As well as a range of smaller species, including methanol (CH_3OH), formaldehyde (H_2CO), and formic acid ($HCOOH$), it also contains some of the largest known COMs, including methyl formate ($HCOOCH_3$), dimethyl ether (CH_3OCH_3), acetone (CH_3COCH_3), ethylene glycol (($CH_2OH)_2$), glycol aldehyde (CH_2OHCHO), vinyl cyanide (CH_2CHCN), ethyl cyanide (CH_3CH_2CN), and acetic acid (CH_3COOH). The more abundant of these species, such as methyl formate, may have a fractional abundance as large as 10^{-7}. The richness of this chemistry implies that many further COM species remain to be discovered.

Tracer molecules: All complex organic molecules provide constraints on the nature of the ice; mantle processing; the frequency of the mantle return mechanism, that is, shocks of modest velocity; and the subsequent gas phase chemistry. All of these issues are currently uncertain.

5.7 Further Reading

Astronomy & Astrophysics. 2010. Special Issue, 518. Herschel: The first science highlights.

Astronomy & Astrophysics. 2010. Special Issue, 521. Herschel/HIFI: First science highlights.

Bergin E., and Tafalla, M. 2007. Cold dark clouds: The initial conditions for star formation. *Annual Review of Astronomy and Astrophysics*, 45, 339.

Bachiller R. 1996. Bipolar molecular outflows from young stars and protostars. *Annual Review of Astronomy and Astrophysics*, 34, 111.

Churchwell, E. 2002. Ultra-compact HII regions and massive star formation. *Annual Review of Astronomy and Astrophysics*, 40, 27.

Draine, B. T. 2003. Interstellar dust grains. *Annual Review of Astronomy and Astrophysics*, 41, 241.

Evans, N. J., II. 1999. Physical conditions in regions of star formation. *Annual Review of Astronomy and Astrophysics*, 37, 311.

Glassgold, A. E. 1996. Diffuse atomic and molecular clouds. *Annual Review of Astronomy and Astrophysics*, 34, 241.

Herbst, E., and van Dishoeck, E. F. 2009. Complex organic interstellar molecules. *Annual Review of Astronomy and Astrophysics*, 47, 427.

Snow, T. P., and McCall, B. J. 2006. Diffuse atomic and molecular clouds. *Annual Review of Astronomy and Astrophysics*, 44, 367.

van Dishoeck, E. 2004. ISO spectroscopy of gas and dust: from molecular clouds to protoplanetary disks. *Annual Review of Astronomy and Astrophysics*, 42, 119.

van Dishoeck, E., and Blake G. 1997. Chemical evolution of star-forming regions. *Annual Review of Astronomy and Astrophysics*, 36, 317.

6

Molecular Tracers in External Galaxies

The first detections of molecular emissions from species such as CO and HCN in external but relatively nearby galaxies were made in the 1970s. From the 1990s, detections were made of CO emission from high-redshift objects, culminating in the remarkable identification in 2003 of high-excitation CO in a gravitationally lensed quasar at a redshift of 6.4. At that time, this was the most distant quasar known. In standard cosmology a redshift of 6.4 represents material when the Universe was merely a few percent of its present age. The discovery of molecular emission in such a distant object demonstrated that chemistry was occurring very early in the evolution of the Universe, and its products – molecules – must therefore also be widespread. In more recent years, it has been firmly established that chemistry in external galaxies can be complex and well developed. So the question arises: Can we use molecular emissions from distant galaxies to explore the physical conditions in them and their likely evolutionary status, as we can do for various regions of the Milky Way (cf. Chapter 5)?

There are a couple of points to be borne in mind, before simply applying the ideas in Chapter 5 to external galaxies. The most important one is that – apart from the nearest objects – most galaxies will be spatially unresolved; the telescope beam will usually encompass the entire galaxy being observed, so that emissions from many types of region are compounded. However, the detection of, say, CO, SiO, and CH_3OH emissions in a spatially unresolved galaxy does not mean that they occur in the same region of that galaxy: the first molecule may indicate the presence of cold tenuous clouds, the second strong shocks, and the third dense star-forming cores. External galaxies will, in general, contain the variety of regions and sources similar to those that we can identify in the Milky Way.

The second point is that galaxies are found in a variety of shapes, sizes, and physical conditions. The range of physical parameters (gas densities, UV fields, cosmic ray ionisation rates, dust properties, etc.) that determine the appearance of the Milky Way may be very different in other galaxies. Therefore, the sensitivity of astrochemistry to those parameters – as discussed in Chapter 4 – is probably important. We should not be tempted to use information about the Milky Way as a reliable guide to the properties of other galaxies, without careful reference to those sensitivities. In particular, the X-factor that was defined for the Milky Way to convert CO integrated line intensities to H_2 column densities (see Section 5.1.2) may give misleading results if applied to other galaxies.

What are the questions we might hope to address in using molecular emission to study galaxies? For the Milky Way (Chapter 5), the main topics for which molecules are ideal probes are the following: tracing the matter that is the reservoir for star formation; describing the process of star formation itself; and determining the influence of newly formed stars on their environments. For galaxies, those questions still apply, but in addition there are even larger questions: How do galaxies form, evolve, and interact with each other, and what are the main power sources for their emissions?

There is, naturally, great interest in the most highly luminous galaxies. For example, *starburst galaxies* are powered by exceptionally high rates of massive star formation, possibly triggered by mergers between galaxies rich in interstellar matter. Such events must be relatively short-lived, persisting only until the interstellar gas reservoir is significantly depleted. Where the interstellar medium is dust-rich, much of the ultraviolet radiation from the massive stars is absorbed by the dust and re-radiated in the infrared, giving rise to a separate classification of *Ultra–Luminous InfraRed Galaxies*, or *ULIRGs*, which, as their name implies, are some of the most powerful infrared sources in the Universe. Other very important galaxies are the *active galaxies*, which have at their centres compact regions that are highly luminous in all wavebands. These central regions are powered by the accretion of mass by supermassive black holes at the centre of the galaxies. Of course, these (and other) classifications of galaxies may not be mutually exclusive. For example, a ULIRG may be re-emitting radiation in the infrared that originated as ultraviolet not only from massive stars in the galaxy but also from an *active galactic nucleus (AGN)* at its centre. In fact, the classification of spatially unresolved galaxies as (for example) starbursts or AGN-dominated is currently controversial. Table 6.1 lists some common types of galaxies and gives their characteristic features.

Fortunately, the chemistry in galaxies is strongly influenced by the local environment. This provides an opportunity to use the signature chemistry arising in different environments to disentangle the molecular line emissions and to identify different components within a galaxy – even though those components are spatially unresolved. The purpose of this chapter is to discuss how

Table 6.1. *Common types of galaxies and their physical characteristics*

Galaxy type	Physical characteristics
Normal spiral	Highly flattened disks supported by rotation, with spiral arms (where star formation occurs) plus central bulge (devoid of young stars); 10^9–10^{12} $M_\odot$; 6–100 kpc; 10^8–10^{11} $L_\odot$
Early-type	Very little gas and dust; mainly populated by old stars; 10^6–10^{13} $M_\odot$; 1–150 kpc; 10^6–10^{12} $L_\odot$
Barred spirals	Spiral galaxies with spiral arms of gas, dust, and stars originating at the ends of a bar passing through the nucleus; similar masses and luminosities as normal spirals.
Irregulars	Not dominated by disk or bulge; 10–20% of their mass is gas; 10^6–10^{11} $M_\odot$; diameter: 1–10 kpc; 10^6–10^9 $L_\odot$
Starbursts	Large reservoirs of gas; high rate of star formation; can be regular or irregular in shape; 10^6–10^{10} $M_\odot$; 100–1000 kpc; 10^9–10^{14} $L_\odot$
Ultra-luminous infrared	Very dusty, possibly hosting powerful AGNs; $>10^{12}$ $L_\odot$
Dwarfs	Very faint galaxies; $\sim 10^6$ $M_\odot$; ~ 1000 pc

this 'disentangling' may be done. In the next section, we describe molecular signatures arising from the various components of a galaxy, and in Section 6.2 we summarise some of the deductions that have been made from molecular line observations of various kinds of galaxies. In Section 6.3, we consider a more coordinated approach to understanding the nature of galaxies by using molecular line emissions.

6.1 Multicomponent Galaxies

Although the nearest galaxies, such as the starburst galaxy M82 (Figure 6.1), can be mapped with single-dish telescopes at a spatial resolution of a few arcseconds, a single-dish beam will necessarily encompass many different gas components. For example, although we can resolve the nucleus of M82, we cannot trace, at any wavelength, the spatial extent of individual star-forming regions, using single-dish telescopes. The Atacama Large Millimeter/submillimeter Array (ALMA) (which with the full array in its most extended configuration will have a resolution of $\lesssim 0.01''$ arcsecs at wavelengths ~ 0.3 mm) will be

(a)

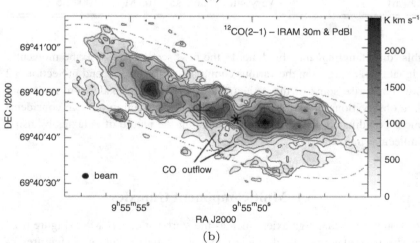

(b)

Figure 6.1. (a) Hubble image of M82, a galaxy with a superwind (Courtesy of
NASA, ESA, The Hubble Heritage Team, STScI/AURA; M. Mountain, P. Puxley,
J. Gallagher). (b) CO(2–1) map of M 82. (Reproduced with permission from Weiss,
A, Neininger, N., Hüttemeister, S., and Klein, U. 2001. *Astronomy & Astrophysics*,
365, 571.) Copyright ESO.

Table 6.2. Example of molecules detected in NGC253 and their fractional abundances with respect to molecular hydrogen

Molecule	Fractional abundance	Molecule	Fractional abundance
$HN^{13}C$	2.51×10^{-11}	HCO^+	1.58×10^{-9}
$H^{13}CO^+$	3.98×10^{-11}	$HNCO$	1.58×10^{-9}
$H^{13}CN$	1.26×10^{-10}	H_2CO	2.51×10^{-9}
SiO	1.26×10^{-10}	OCS	3.80×10^{-9}
CH_3CN	3.16×10^{-10}	CN	5.01×10^{-9}
$C^{34}S$	3.98×10^{-10}	HCN	5.01×10^{-9}
$c\text{-}C_3H_2$	5.01×10^{-10}	CS	6.31×10^{-9}
HC_3N	6.31×10^{-10}	CH_3CCH	6.31×10^{-9}
N_2H^+	6.31×10^{-10}	CH_3OH	1.26×10^{-8}
HNC	1.00×10^{-9}	C_2H	2.00×10^{-8}
SO	1.26×10^{-9}	NH_3	6.31×10^{-8}

Data adapted from Table 7 in Martin, S., Henkel, C., Garcia-Burillo, S. 2006. *Astrophysical Journal Supplement Series*, 164, 450.

able to resolve structures down to 0.15 pc at the distance of M82 (3.3 Mpc). Although this is certainly smaller than a typical star-forming region it is larger than an individual star-forming core.

We now have a wealth of molecular data for at least the nearest galaxies and a rapidly growing array of information on distant unresolved galaxies. What are these molecular detections telling us? Certainly, the chemical diversity and complexity that we find indicate that the molecular emission is not all coming from the same component. For example, an unbiased molecular line survey in the wavelength range 1.7–2.3 mm towards the nuclear region of the starburst galaxy NGC253 shows the presence of more than 20 different molecular species (see Table 6.2). Such chemical complexity cannot be explained by a one-component model, and indicates how relative abundances between molecules may be able to provide insights into the physical distribution of the molecular gas in these galaxies. Determining the physical structure of galaxies via the use of molecular emissions may even be more powerful at high redshift, where the beam encompasses the whole galaxy. For example, observations of APM08279+5255, a galaxy at a redshift of $z = 3.911$, reveal ratios of molecular column densities such as HCN/HCO^+ and HCN/CO to be very different from those found in the Milky Way or in nearby galaxies (see Table 6.3). These differences have been interpreted as anomalous and an indication either of high ionisation rates or of high star-formation rates. On the other hand, it is likely that

Table 6.3. HCN/HCO$^+$ and HCN/CO in the Milky Way, a nearby
galaxy (NGC1068), and a high-redshift galaxy (APM08279+5255)

	Milky Way	NGC1068	APM08279+5255
HCN/HCO$^+$	1–5	2.5	10
HCN/CO	1×10^{-4}	3×10^{-4}	1×10^{-2}

multiple components, each with different physical conditions, are contributing
to the emission, so that a knowledge of the individual components is required
to interpret such ratios. Molecular emissions can give information about those
individual components.

Astrochemistry tells us that the detection of species such as HCO$^+$ and
CH$_3$OH in the same observation of an unresolved galaxy implies that emission
is coming from different interstellar components with different physical conditions. We know from Chapters 3 and 4 how the most common interstellar
molecules form, under what conditions they become abundant, and therefore
which physical parameters they trace. Which molecules should we use to trace
physical conditions in galaxies? Can the measured abundances tell us if some
galaxies may be dominated by one or more type of energy source? We approach
these questions by considering, first of all, those galaxies that appear to be dominated by a single energy source.

6.1.1 Photon-Dominated Regions

The radiation emitted by some galaxies appears to be dominated by emission
from giant Photon-Dominated Regions (PDRs) that are, of course, powerful
emitters in molecular lines in the millimetre and submillimetre regions of the
spectrum. Giant PDRs may form in galaxies where clusters of young massive
stars are present and in galaxies dominated by radiation coming from an active
nucleus. In this section we discuss the chemistry of such galaxies and therefore identify the molecular signatures one should expect to find in a galaxy
dominated by giant PDRs.

Whether a molecule is a good tracer of an extragalactic PDR depends,
in most cases, on the physical and chemical conditions of the galaxy. The
metallicity of a galaxy can of course greatly affect its chemical composition
(see Section 4.3): as discussed earlier, very low values of metallicity imply
very low abundances for most species (with a few exceptions such as CO,

H_2O, HCO^+, H_3O^+, and OH), and in such regimes deuterated ions may also be more abundant than expected. High metallicities (compared to the Solar System) lead to complex trends, with species falling into one of the following categories: some species have relative abundances varying linearly with the metallicity changes; other molecules show chemical abundances that are rather insensitive to the metallicity changes, whereas some species have abundances that are inversely dependent on metallicity (see Table 4.1).

External galaxies may differ from the Milky Way not only in metallicity but also in their initial chemical composition. For example, there is evidence that second-generation galaxies had a reduced nitrogen elemental abundance, whereas both sulfur and oxygen may be enhanced. Sulfur-bearing species are very sensitive to the elemental abundances in a galaxy.

The strength of the radiation field also affects the chemical composition of the galaxy: in general, high radiation fields tend to suppress the chemistry, with some species, such as CS, OCS, and SO_2, being highly sensitive to the UV intensity. Some others (e.g., C_2 and C_2H) even show a small increase in their abundances when the far UV intensity is high, owing to the injection of free carbon from the photodestruction of CO. Of course, variations in density and cosmic ray ionisation also influence the chemistry of PDRs. We discuss these effects in the next sections.

In the following paragraphs we list selected molecular species that, according to chemical models, are useful tracers of PDRs:

HCO^+: This molecule is abundant in PDRs regardless of the metallicity, oxygen or sulfur initial elemental values, cosmic ray ionisation rates, far UV radiation field, and gas density of the galaxy. Its estimated fractional abundance, between 10^{-12} and a few times 10^{-11}, is low; however, this is one of the very few species that can be defined as a true tracer of PDRs in galaxies because of its insensitivity to physical changes (as long as the cosmic ray ionisation rate is not higher than $\sim 10^{-15}$ s^{-1}).

H_3O^+: This molecule, similar in abundance to HCO^+, is another ion that is generally enhanced in PDRs and is fairly insensitive to the initial conditions. However, of particular interest is the fact that a very low metallicity, while deleterious for most molecules, seems to enhance H_3O^+; this enhancement is caused by a severe reduction in the electron density at low metallicity. Because the main loss of polyatomic ions is through dissociative recombination, the ion abundance increases.

CO: This molecule is of course very abundant in PDRs, and even at low metallicities, where its fractional abundances may decline up to a factor of 1000 from its canonical value of $\sim 10^{-4}$, it is the most abundant species in PDRs. As noted earlier, this molecule is ubiquitous, so that its usefulness is still mainly to trace the molecular gas rather than a particular process or component within a galaxy. On the other hand, it is the only molecule for which the complete J-ladder can be observed for a variety of extragalactic sources, making CO a powerful tracer of the excitation conditions of a galaxy.

H_2O **and CS**: At high visual extinctions ($A_V \sim 4$), these two species are abundant in PDRs regardless of the physical conditions of the galaxy. However, for a cloud at such extinctions to be a PDR requires an intense far UV radiation field. Hence, these molecules trace those PDRs in which the radiation field is very intense.

HCO: This molecule has also been observed to be a particularly good tracer of the PDR interfaces. The ratio [HCO^+]/[HCO] has been found to be $\sim$2.5–30 in prototypical galactic PDRs, and this, together with a large ($\gtrsim 10^{-10}$) HCO abundance, is believed to be a diagnostic for an ongoing far UV-dominated photochemistry.

CO^+: This molecule is formed by direct photoionisation of CO where the PDR is driven by intense and hard radiation fields. High abundances of this molecule appear to be correlated to similar enhancements of HOC^+, and [CO^+]/[HOC^+] ratios in the range of 1–10 are observed in a number of PDRs.

6.1.2 Cosmic Ray–Dominated Regions

The generation of cosmic rays is believed to be associated with the formation of massive stars. Therefore, some galaxies showing starbursts or undergoing galaxy mergers may have exceptionally high cosmic rays fluxes. Galaxies (or parts of them) with these properties are termed cosmic ray–dominated regions (CRDRs). The effects of high levels of cosmic ray fluxes on the molecular and atomic gas were briefly discussed in Section 4.2. High fluxes will raise both the kinetic temperature of interstellar gas and the ionisation fraction even in UV-shielded regions.

Deviations from the Milky Way value of the cosmic ray ionisation rate, ζ, can lead to a very diverse chemistry. As discussed in Section 4.2, up to a critical value of the ionisation rate, $\sim 10^{-15}$ s^{-1}, the chemistry should be as rich as in the Milky Way Galaxy (see Figure 4.2). However, if ζ is increased to about 10^{-12} s^{-1} much of the chemistry is suppressed because – for such a high rate – molecular hydrogen, which is the key to all interstellar chemistry, would be destroyed faster than it can be replaced, and the rise in temperature driven by cosmic ray energy dumped into the gas (see Figure 6.2) means that

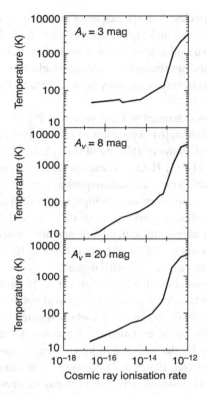

Figure 6.2. Steady-state gas temperatures determined self-consistently with chemical evolution at depths corresponding to $A_V = 3$, 8, and 20 magnitudes into a cloud of uniform number density of H nuclei, 10^4 cm^{-3}. The cosmic ray ionisation rate per second is indicated on the abscissa. The cloud has a metallicity of $0.1 \times$ solar and a gas dust mass ratio of 1000. (Reproduced with permission from Bayet, E., Williams, D. A., Hartquist, T. W., and Viti, S. 2011. *Monthly Notices of the Royal Astronomical Society*, 414, 1583.)

reactions with hot atomic hydrogen destroy all other molecules. It is unclear how high the cosmic ray energy densities may become, though we do know that star formation rates in some galaxies may be several orders of magnitude larger than in the Milky Way. If cosmic ray fluxes are enhanced by up to 10^4 times the Milky Way value ($\sim 10^{-17}$ s^{-1}) then a diverse and useful chemistry should still exist and could be used to trace CRDRs. Nevertheless, it is true that most molecular species decrease with an increase in cosmic ray ionisation rate; in the paragraphs that follow we identify some molecules that could be particularly useful when tracing CRDRs in external galaxies.

Sulfur-bearing species: The species most sensitive to the cosmic ray ionisation rate are OCS, SO_2, and H_2CS, all species that, to date, have not been widely observed in extragalactic environments. In fact, the predicted fractional abundances of these species relative to hydrogen decline, from $\sim 10^{-7}$–10^{-9} for Milky Way conditions to values that may be undetectable ($\sim 10^{-12}$) for models with ζ of the order of 10^{-15} s^{-1}.

C_2: This species is enhanced at high values of ζ owing to the increase in C^+ released in the dissociative ionisation of CO by He^+, and peaks for ζ of $\sim 10^{-13}$ s^{-1}, regardless of the metallicity of the galaxy.

H_2O, OH, OH^+, H_2O^+, H_3O^+: These species are in general enhanced at high cosmic ray ionisation rates and tend to peak at $\zeta \sim 10^{-14}$ s^{-1}; some species such as OH and OH^+ may even sustain high values of fractional abundances ($\sim 10^{-7}$–10^{-8}) for values of ζ approaching 10^{-12} s^{-1}. H_3O^+ is of particular interest as it becomes abundant (with a fractional abundance larger than 10^{-10} relative to hydrogen) only for values of ζ of the order of 10^{-13} s^{-1} but it declines again at even higher cosmic ray ionisation rates.

HCO^+: Routinely used as a tracer of cosmic rays in our own galaxy, HCO^+ is a useful cosmic ray tracer for values of ζ up to about 10^{-13} s^{-1}.

The behaviour of these oxygen and carbon species is a consequence of the thermal stimulation of the endothermic reactions that may initiate oxygen and carbon chemistry by the increase in kinetic temperature arising from the higher ionisation rates (see Figure 6.2). However, ultimately, the reduction in the H_2 fraction and the destruction by hot H atoms suppress all these reaction pathways.

Some extragalactic environments may host a high cosmic ray ionisation rate at large (greater than solar) metallicities, as may be the case for ULIRGs. It is worth noting, therefore, that the choice of potential molecular tracers will depend heavily on the metallicity of the galaxy; in CRDRs the metallicity determines the abundance of C^+, one of the most important drivers of cosmic ray–dominated chemistry.

6.1.3 X-ray–Dominated Regions

X-ray emission is particularly important for galaxies containing active nuclei (AGNs). The molecular gas in these environments can be exposed to strong X-ray irradiation. Unlike UV photons, which are absorbed by dust in path lengths of a few visual magnitudes, X-rays can penetrate gas column densities as large as $N(H_2) \sim 10^{23}$–10^{24} cm^{-2} before being significantly attenuated (see Section 3.2.2). Therefore, X-ray–dominated regions (XDRs) can be the dominant sources of emission from molecular gas in the vicinity of AGNs. In fact,

XDRs are often invoked to explain so-called 'exotic' or 'anomalous' molecular line ratios (such as the large HCN/CO abundances measured in the Seyfert 2 galaxy NGC1068).

Somewhat more direct evidence for XDR chemistry comes from the high fractional abundances of SiO ($\sim 10^{-9}$) observed in galaxies such as NGC1068. However, strong shocks may also be contributing. In fact, it is difficult to distinguish CRDR chemistry from that of XDRs; species such as CN, HCO, HCO^+, and HOC^+ have all been used as indicators of XDRs. Nevertheless, line intensities and abundance ratios may vary between the two radiation sources. For example, chemical and radiative transfer models show that CO rotational line intensity ratios increase when cosmic ray ionisation rates are enhanced, but they remain smaller than those in X-ray–dominated regions. In particular, high-J ($J > 10$) CO lines may allow us to distinguish between PDRs with enhanced cosmic ray ionisation rates and XDRs.

6.1.4 Molecular Clouds and Dense Star-Forming Gas

As we have seen in Chapter 1, Giant Molecular Clouds (GMCs) are the sites of most of the star formation that takes place in galaxies. In external galaxies star-forming clumps will be embedded, and observationally unresolved, in GMCs. The gas in GMCs will have average number densities of the order of $\sim 10^3$ cm^{-3} and temperatures of 50–100 K, while the star-forming clumps may have number densities up to $\sim 10^7$ cm^{-3} and temperatures as low as 10 K. Hence, unresolved by observations, extragalactic gas will span large ranges of density and temperature. This issue becomes particularly important when estimating the molecular mass of a galaxy from observations of CO, via the X-factor (see Section 5.1.2). Table 6.4 shows values of X computed for several different galaxy types and several different spectral lines.

Molecular observations are an ideal tool to trace a wide range of densities in the interstellar medium of a galaxy, largely because of the wide range of critical densities among different molecular species and among the transitions of the same molecules (see Chapter 10). Chapter 5 describes a variety of molecules that can be used to trace resolved regions in the Milky Way differing in density. However, in extragalactic environments density structures are very rarely resolved, so that the observer is left with the apparently intractable problem of interpreting molecular observations assuming that the gas density has some average value. How can we identify unresolved structures of high-density gas?

The molecules most often used to demonstrate the presence of dense gas in external galaxies are HCN and HNC. However, regions of high-density

Table 6.4. Variation of the X-factor (in units of 10^{20} (K km s^{-1})$^{-1}$) for several galaxy types for several J transitions of CO, as well as for the [C I] 609 μm transition

Galaxy	$X(1-0)$	$X(2-1)$	$X(3-2)$	$X(4-3)$	$X(6-5)$	$X(9-8)$	$X($[C I] 609 μm)
Normal spiral	0.3	0.2	0.2	0.4	1.2	4.2	0.6
Starburst	2.0	2.1	3.5	10.4	25.9	61.2	1.2
Irregular starbust	1.5	1.7	2.9	9.5	79.4	203.7	1.3
Dwarf irregular	9.5	12.0	37.7	179.3	203.9	508.0	1.9

Values taken from Bell, T. A., Viti S., and Williams, D. A. 2007. *Monthly Notices of the Royal Astronomical Society*, 378, 983.

gas are not necessarily hosting star formation. It is clear that variations in density significantly affect the molecular content of a galaxy, but the degeneracy between density and temperature implies that single transitions of one molecule should not be used to determine quantitatively the gas density, especially in extragalactic environments. Recent models and observations indicate that one way to trace different density structures is by the use of multiline multispecies observations.

In the paragraphs below we list some species and transitions that may be useful when trying to disentangle the high-density components of extragalactic gas. However, we note that this list is not comprehensive because of the very limited number of molecular observations in extragalactic environments that have been made.

CS: This molecule is well known for tracing dense clumps in GMCs in the Milky Way. Recent multiline surveys of this molecule in extragalactic environments have revealed low J (up to 4) transitions of CS trace gas densities of the order of 10^5 cm^{-3}, while high J (up to 7) transitions of CS trace gas of at least an order of magnitude higher density, possibly therefore tracing star-forming gas.

HCN and HNC: These two species have routinely been used as tracers of star-forming gas in external galaxies. In fact, HCN seems to trace the same densities as the low-J CS (and hence not necessarily star-forming gas), whereas HNC peaks at higher densities and may therefore be a better tracer of high-density star-forming gas.

CH$_3$OH: The methanol 'ladder', like that of CS, is a useful diagnostic of multiple density components. However, because methanol is also a good tracer of shocks, and hence of high temperatures, it is often difficult to resolve the degeneracy between temperature and density when extrapolating physical parameters from methanol observations.

6.2 Characterisation of Galaxies via Molecules

So far in this chapter, we have highlighted those molecules already detected in galaxies and those that are predicted by chemical models to be useful tracers in situations where a single physical parameter is enhanced. For example, galaxies dominated by PDRs are obviously assumed to have a very intense UV radiation field, and some useful tracers of these galaxies are listed in Section 6.1.1. Galaxies in which CRDRs are important obviously have very high cosmic ray fluxes, and some tracers of those situations are listed in Section 6.1.2. In these chemical models, however, it is generally assumed that parameters other than the one that is enhanced take some particular canonical values, such as those of the Milky Way Galaxy.

Although this approach is useful in identifying important molecular tracers of galaxies with a single particularly enhanced physical characteristic, it is also an oversimplification. For example, galaxies with intense UV fields may contain many massive stars to generate those fields, and these massive stars are also likely to generate intense fluxes of cosmic rays. Therefore, galaxies with enhanced UV fields are also likely to have enhanced cosmic ray fluxes. It is necessary, therefore, to consider predictions from more realistic models in which the physical parameters adopted (radiation field intensity, cosmic ray flux, density, dust to gas ratio, etc.) are given values that best represent particular galaxy types. In this section, therefore, we consider several more realistic galaxy types that are identified by their morphology and by the physical processes that characterise them and discuss the molecular tracers of their physical conditions. In reality, of course, the galaxy types are probably not totally distinct but merge into one another.

6.2.1 Molecular Tracers in Galaxies with Intense UV Fields

Here, we consider the predictions of chemical models for three types of galaxy. Molecular clouds in starburst (SB) galaxies such as M83 and NGC 253 can be rather crudely represented in chemical models with values of metallicity, dust to gas ratio, H_2 formation rate, cosmic ray ionisation rate, and elemental abundance ratios near to Milky Way values. The clouds may be represented by hydrogen number densities of 10^4 cm^{-3} and high visual extinctions, A_V, of about 8 magnitudes. However, the UV radiation field intensity impinging on these clouds is expected to be very high, up to about 10^5 times that of the Milky Way.

Other galaxies, such as Arp 220 and NGC 3079, seem to have aspects of starburst characteristics while also possessing active nuclei. These SB+AGN galaxies may be represented in chemical models by a set of parameters similar to those for SB galaxies, but with a UV radiation intensity lower by a factor of 100 (due to their lower star formation rate) and a cosmic ray ionisation rate higher by a factor of 100 (due to AGN activity).

At high redshift, the properties of galaxies are even more uncertain, but for APM 08279 (a gravitationally lensed quasar at a redshift of 3.91) and the Cloverleaf (or H1413+117, a gravitationally lensed quasar at redshift 2.56) various molecules have been detected in both dense and less dense gas. Such galaxies may be crudely represented by chemical models with metallicity, dust to gas ratio, H_2 formation rate, and elemental abundance ratios reduced by a factor of 10 compared to the Milky Way, with a high UV field ($\sim 10^5$ times the Milky Way value), and a high cosmic ray ionisation rate ($\sim 10^{-15}$ s^{-1}).

The results of these crude chemical models are provided in Table 6.5, which shows whether molecules are likely to be detectable (for an assumed detectability limit for a fractional abundance of 10^{-12}). Evidently, there are many possible tracers of PDRs in these galaxy types; of these, CO and H_2O are – not surprisingly – very abundant, whereas CS, CN, and OH have lower abundances but should still be potential tracers. These studies also predict that HCN, HNC, and HCO$^+$ – often used as molecular tracers of dense gas in galaxies – have rather low abundances. Evidently, these crude models suggest that there are many so far untested potential molecular tracers of galactic PDRs.

6.2.2 Hot Core Molecules in Galaxies

The formation of massive stars requires the formation of dense hot cores and of the characteristic (and relatively large) molecules they contain (see Section 5.4.1). These signature molecules are created in the rich solid-state chemistry occurring in the interstellar mantles on dust grains and released as the ices warm up during the star formation process. The galactic interstellar radiation field is unimportant for hot core chemistry because hot cores have very high visual extinction. These hot core molecules may be regarded as good tracers of the formation of massive stars in galaxies. Hot cores arise in galaxies in which massive star formation occurs, from normal galaxies to starbursts. In particular, the galaxies do not need to be spatially resolved; the integrated emission from

Table 6.5. Detectability of 19 species likely to trace the dense PDR gas component in three categories of galaxies

Molecule	Starburst	Starburst + AGN	High redshift
CO	+	+	+
H_2O	+	+	+
CS	+	+	+
SO	+	+	+
CN	+	+	+
OH	+	+	+
HNC	+	+	+
HCN	+	+	+
HCO^+	+	+	+
H_3O^+	+	+	+
C_2	+	+	+
C_2H	+	+	+
CO_2	+	+	+
OCS	+	−	−
SO_2	−	−	−
H_2S	−	−	−
H_2CS	+	+	−
H_2CO	+	+	−
CH_2CO	+	−	−

The symbol (+) signifies likely detectability, while the symbol (−) indicates probable non-detectability. Reproduction of Table 10 from Bayet, E., Viti, S., Williams, D. A., Rawlings, J. M. C., and Bell. T. A. 2009. *Astrophysical Journal*, 696, 1466, with permission.

hot core molecules gives a direct and independent measure of the instantaneous star formation rate, so this technique may be particularly useful for starburst galaxies at high redshift. The main uncertainty is in the number of hot cores in the observed galaxy. The Milky Way Galaxy is estimated to have about 10^4 hot cores, and its average star formation rate is about 1 solar mass per year. Distant galaxies have star formation rates several orders of magnitude larger, so if the number of hot cores is proportional to the star formation rate, then these galaxies may contain 10^7 or more hot cores.

In this subsection, we discuss the predictions of chemical modelling of the hot core chemistry for several crude models of galaxies. We represent typical spirals (such as IC 342 and NGC 4736) by a model with Milky Way parameters, including a hot core temperature of 300 K. A starburst galaxy (such as M83) may be expected to have a higher temperature in the hot cores, and we adopt 500 K in the starburst model.

Some galaxies (such as IC 10) are known to have low metallicity. For this model we adopt a metallicity lower than in the Milky Way by a factor of 5 and a hot core temperature of 500 K, as IC 10 is marginally starburst; otherwise we use Milky Way parameters.

Finally, we consider a model intended to represent hot core chemistry in a high-redshift galaxy such as the Cloverleaf or APM 08279. The physical parameters of these galaxies are not known, but considering its redshift we will assume a metallicity and dust: gas ratio depleted by a factor of $\sim$20, a cosmic ray ionisation rate enhanced by a factor of 10, with respect to the Milky Way, and a hot core temperature of 500 K.

Of course, these models are only representative and should not be considered as an accurate model of any particular source.

The results from these crude models are summarised in Table 6.6, which lists the molecules likely to be detected in each galaxy type, on the assumption of a limit of detectability for a fractional abundance of 10^{-12}. These computations suggest that there should be some differences between the tracers with respect to the various galaxy types (although it may be difficult to distinguish easily between hot core molecules in the normal spiral and the starburst galaxies). Even in the case of the high-redshift model in which the metallicity is heavily depleted, some hot core molecules should have fractional abundances easily capable of detection (e.g., HCN, HNC, CS, H_2CS, and CH_3CN). We conclude that many potential molecular tracers of hot cores in external galaxies remain to be studied.

6.2.3 Disentangling Galactic Components

Much of the mass of interstellar gas in galaxies is in PDRs. Yet Table 6.5 indicates that different galaxy types should have different molecular tracers of PDRs, and it should therefore be possible to distinguish between them. In addition, the inclusion of a new energy source such as an AGN in a galaxy model changes the range of molecular tracers that can be identified. The star formation rate in a galaxy, whether that galaxy is resolved or not, can be addressed through the emissions from the hot core molecules generated during star formation. Thus, there is a real possibility of disentangling the information in molecular line spectra of galaxies and of inferring the nature of the energy sources and other processes driving the evolution of the galaxy. This has, so far, been done only for a few nearby galaxies, such as NGC253, M82, and NGC1068, as we see in Sections 6.3.3 and 6.3.4.

Table 6.6. Detectable tracers of extragalactic star-forming chemistry for four examples of models likely to be appropriate for representing four kinds of galaxies

Molecule	Spiral	Starburst	Low metallicity	High redshift
CS	+	+	+	+
OCS	+	+	+	−
CH_3CN	+	+	+	+
CH_3OH	+	+	+	−
SO	+	+	+	+
SO_2	+	+	+	−
H_2S	+	+	+	−
H_2CS	+	+	+	+
H_2CO	+	+	+	+
CH_2CO	+	+	−	−
C_2H_5OH	−	−	−	−
$HCOOCH_3$	+	+	−	−
HNC	+	+	+	+
HCN	+	+	+	+
HCO^+	−	−	−	+

The symbol (+) signifies likely detectability, while the symbol (−) indicates probable nondetectability. Reproduction of Table 9 from Bayet, E., Viti, S., Williams. D. A., and Rawlings, J. M. C. 2008. *Astrophysical Journal*, 676, 978, with permission.

6.3 Recent Molecular Line Studies of Galaxies

6.3.1 Normal Spiral Galaxies

A spiral galaxy is a flat, rotating, and relatively thin disk of stars, gas, and dust; it has spiral arms in the plane of the disk extending out from the centre, together with a concentration of stars towards the centre – the so-called bulge. This complex structure is surrounded by a much fainter spheroidal halo of stars, including stars in globular clusters. Most of the star formation in a spiral galaxy occurs in the spiral arms, which are the locations of the giant molecular clouds. These arms can be traced in nearby galaxies by the millimetre-wave emission from the giant molecular clouds, by optical emission from the young massive stars and the HII regions they generate, and from infrared emission from dust heated by stars deeply embedded in dark clouds. The spiral arms in many spiral galaxies are traced from the ends of a bar-like structure that bisects the central bulge. The Milky Way Galaxy is, of course, an example of a barred spiral galaxy.

Spiral galaxies provide a variety of environments in which to study the processing of molecular gas. Spiral galaxies have been extensively mapped in the millimetre observations of CO and spirals have been sub-classified according to their molecular content-which, surprisingly, varies by huge amounts even within the same class of massive spirals.

Single-dish CO observations, with relatively low angular resolution, were generally first performed to derive global properties of the molecular gas; its total amount; proportion with respect to HI; and its relation to metallicity, morphology, and star formation. More detailed studies of CO have also been undertaken, mostly with interferometers. As for every other type of galaxy, the standard conversion factor between H_2 and CO that is used to derive molecular masses from CO observations is unreliable, because the conversion factor depends on the local physical conditions (see Figure 5.1). Nevertheless, CO observations have played a key role in determining the location of the molecular gas in spirals.

There is, however, remarkably little in terms of observational data of other molecular species in spiral galaxies. HCN and HCO^+ have been observed in several giant molecular clouds in M31, and the HCN/HCO^+ ratio is found to be rather higher than that in the Milky Way galaxy but still well below that found in starbursts, indicating that indeed the radiation fields are weaker in spirals than in starbursts, as expected.

6.3.2 Early-Type Galaxies

Early-type galaxies comprise different types of galaxies, in particular, lenticular and elliptical. Until recently they were thought to be the endpoint of galaxy evolution because they have uniform red optical colours and are located in a tight red sequence in an optical colour-magnitude diagram. The separation between early-type galaxies and star-forming galaxies in such diagrams implies that the fuel for star formation in early-type galaxies must be consumed, destroyed, or removed on a reasonably short timescale. Several removal mechanisms have been proposed, all efficient enough to leave these galaxies on the red sequence with little or no cold interstellar gas, and so – in theory – with little or no star formation occurring.

However, observations of neutral hydrogen reservoirs and of dust, as well as UV excess, not attributable to old stellar populations, revealed that some of these galaxies do indeed have cold gas reservoirs and residual star formation. Clearly, molecules are also present in these galaxies and in fact CO observations since the mid-1990s have revealed at least a 30% rate of detection of molecular gas.

A complete, volume-limited survey of the properties of 260 morphologically selected early type galaxies has revealed substantial molecular gas reservoirs (10^7–10^9 $M_\odot$ of H_2) via CO observations. Theoretical studies of the properties of this molecular gas in metal-rich early-type galaxies show that in PDRs the fractional abundances of some species such as CS, H_2S, H_2CS, H_2O, H_3O^+, HCO^+ and H_2CN seem invariant to an increase of metallicity whereas C^+, CO, C_2H, CN, HCN, HNC, and OCS appear to be sensitive. Early-type galaxies are also believed to be enhanced in their α-elements (formed in α-processes; their most abundant isotopes are integer multiples of the mass of the helium nucleus). The most sensitive species to the change in the fractional abundance of α-elements seem to be C^+, C, CN, HCN, HNC, SO, SO_2, H_2O, and CS. These initial studies are excellent examples of preparatory work for ALMA, as they provide line brightness ratios for all these species. It is likely that a more complex chemistry will be revealed by future observations of these galaxies.

6.3.3 Starburst Galaxies

Starburst galaxies are extremely luminous galaxies, powered by bursts of massive star formation. Massive stars, thousands to tens of thousand times more luminous than the Sun, occur in clusters and are embedded in dusty molecular clouds. As the lifetimes of massive stars are relatively short (tens of millions of years), periodical bursts of massive star formation ensure an active and rich chemistry in the surrounding gas.

Starburst events are detected indirectly in a variety of ways, all as a result of the effects of the hot massive stars present in them. We see strong optical emission lines and continuum; starbursts exhibit an infrared excess attributed to the dust (heated by radiation from the hot stars). In the radio regime there is emission from supernovae remnants. There is, in fact, a strong correlation between the thermal far infrared emission and the nonthermal radio emission. That these galaxies are characterised by short but frequent bursts of massive star formation is generally accepted and the total mass of massive (≥ 10 $M_\odot$), short-lived ($\lesssim 2 \times 10^7$ yrs) stars can exceed 10^8 $M_\odot$. What causes the bursts, however, is still debatable, but it is possible that mergers of galaxies or gas flows in barred galaxies are responsible.

Observations of starburst galaxies in the millimetre waveband have detected many molecular species. This is, of course, of particular importance because it means that, via observations of CO and other species, one could – with some assumptions – infer the column density of molecular hydrogen and hence

Table 6.7. Comparison of fractional abundances for the two starburst
galaxies M82 and NGC253, and the Orion Bar in the Milky Way

Molecule	Orion Bar	NGC253	M82
CS	2.5×10^{-8}	6.3×10^{-9}	6.3×10^{-9}
SO	1.0×10^{-8}	1.3×10^{-9}	$<3.2 \times 10^{-9}$
C_2H	2.0×10^{-9}	2.0×10^{-8}	2.5×10^{-8}
CH_3OH	1.0×10^{-9}	1.3×10^{-8}	$<5.0 \times 10^{-9}$
CH_3CN	$<5.0 \times 10^{-11}$	3.2×10^{-10}	2.0×10^{-10}
HOC^+	3.2×10^{-9}	2.6×10^{-9}	4.0×10^{-9}
H_2CO	6.3×10^{-9}	2.5×10^{-9}	6.3×10^{-9}
$c-C_3H_2$	2.0×10^{-10}	5.0×10^{-10}	8.0×10^{-9}

The abundances are taken from Martin, S., Mauersberger, R., Martin-
Pintado, J., Henkel, C., and Garcia-Burillo, S. 2006. *Astrophysical Journal Supplement Series*, 164, 450.

the molecular mass of the galaxy and its star formation rate. Of course, the canonical conversion factor from CO to H_2 will most likely not be valid for starburst galaxies because the high UV radiation fields will affect this ratio (see Figure 5.1). Starburst galaxies are among the most luminous in CO; the CO luminosity is strongly correlated to the far infrared flux. But putting aside (for the time being) the problems inherent with the use of the canonical conversion factor between H_2 and CO and using Milky Way values, studies suggest that nearby starburst galaxies have molecular masses ranging between 10^9 and 10^{10} solar masses and that most of the CO is concentrated towards the centre of the galaxy.

CO is of course not the only molecule observed in starburst galaxies. About 45 molecular species have been detected (by 2012) in starburst galaxies. We use two of the most studied starburst galaxies, NGC 253 and M82, as examples of the variety of molecules detected in these objects and of what they can tell us about the energetics of starbursts.

Both galaxies have been the study of extensive submillimetre spectral surveys, especially in their nuclei. These two galaxies are believed to be classical starburst galaxies, IR-bright ($\sim 3 \times 10^{10}$ $L_\odot$), and with strong radio continuum. They are relatively close to the Milky Way Galaxy (~ 3 Mpc). For both galaxies, the observed abundances show striking similarities with those observed in molecular clouds of the galactic centre of the Milky Way, believed to be dominated by low-velocity shocks (see Table 6.7).

However, at a closer look, the chemistry in NGC253 appears different from that of M82; the chemistry and to a large extent the heating in the central region of NGC253 are believed to be dominated by large-scale low-velocity shocks. The similar chemical composition found in the nuclear region of NGC253 to that in the Milky Way star–forming molecular complexes points to an earlier evolutionary stage of the starburst in NGC253 than that in M82. The abundances of molecules such as SiO, CH_3OH, HNCO, CH_3CN, and NH_3 are systematically lower in M82 than in NGC253, whereas species such as HCO and C_3H_2 are more abundant. This may be explained in terms of PDR chemistry; M82 is often associated with more spatially extended PDRs than NGC253. This explanation is consistent with the fact that M82 has many more HII regions than NGC253: intense UV radiation from massive stars is one of the main mechanisms responsible for the heating of the interstellar medium in the nuclear region of starburst galaxies.

This heating mechanism is particularly important in the later evolutionary stages of starburst galaxies where the newly formed massive star clusters are responsible for creating large PDRs. This is the case for M82, where the observed large abundances of molecular species such as HCO, HOC^+, CO^+, and H_3O^+ are claimed to be probes of the high ionisation rates in large PDRs formed as a consequence of its extended evolved nuclear starburst. Observational evidence points to a significant enhancement in the abundance of HOC^+ in regions with large ionisation fractions. The abundance ratio $[HCO^+]/[HOC^+] = 270$ is found in the prototypical galactic PDRs of the Orion Bar, for example. Similar or even lower abundance ratios are observed in the other galactic and extragalactic PDRs (e.g., NGC7023, Sgr B2(OH), and NGC2024). This is in contrast with the larger ratios of ~ 1000 found in dense molecular clouds well shielded from the UV radiation. However, these low HCO^+/HOC^+ ratios are not found in other galactic PDRs.

6.3.4 AGN-Dominated Galaxies

Some galaxies have powerful AGN in their centres, as well as extended bursts of star formation. For these objects the connection between the starburst phenomenon and the central AGN is not well understood, but the presence of large amounts of circum-nuclear gas implies an interaction between the molecular gas that fuels star formation with centres of the gravitational potentials where the AGN resides.

Objects such as NGC 3079 fall into this category. It is reasonable to assume that such galaxies have an enhanced (possibly by a factor of 100) cosmic ray ionisation rate with respect to normal spiral galaxies, as well as a high-intensity radiation field ($\sim 10^3$ higher than the galactic interstellar radiation field). There is no unique molecular tracer of such complex systems: all the molecules that should be detectable are also characteristic of 'pure' starburst galaxies or normal spirals. Observations seem to indicate that abundance ratios of specific molecules, such as HCN/HCO$^+$, may help to trace the energetic processes that dominate these complex galaxies. However, as explained earlier, the chemistry of these molecules is highly dependent on the physical and chemical conditions of the gas. Although some models predict, for example, that HCO$^+$ is enhanced in XDRs, others predict the opposite. Ultimately, species such as HCO$^+$ are highly sensitive to parameter changes and, as shown in previous sections, can vary in abundance by several orders of magnitude during the chemical evolution or because of variations in the cosmic ray and starlight ionisation and UV field rates.

6.3.5 Ultra-Luminous Infrared Galaxies

Ultra-Luminous Infrared Galaxies (ULIRGs) are generally very dusty objects, in which the ultraviolet radiation produced by the obscured protostars is absorbed by the dust and re-radiated in the infrared at wavelengths of $\sim 100~\mu$m. These objects can be more than 100 times more luminous in the infrared than in the optical. ULIRGs are powered by AGNs as well as starbursts; molecular studies of their interstellar gas indeed show signatures of both. For example, the presence of copious amounts of molecular gas has been inferred from CO data ($> 10^{10}$ M$_\odot$), leading to values for the star formation efficiency (defined as the ratio between far-infrared luminosity and molecular gas mass) of 20–200 L$_\odot$/M$_\odot$ (compared to the 4 L$_\odot$/M$_\odot$ in normal spiral galaxies). In fact, these values are larger than those found in some starburst galaxies such as M82. On the other hand, observations of PAHs and silicate dust emissions indicate the presence of (sometimes 'buried') AGN. Whether ULIRGs are dominated by starburst or AGN is still under debate.

Arp 220 is regarded as the prototypical ultraluminous galaxy. Interferometric images of the CO(1–0) emission and also HCN and HCO$^+$ emission show that 90% of the dynamical mass of the system and 75% of the molecular mass is confined to a central core, 600 pc in diameter. The CO (3–2) distribution also shows that there are three dense peaks, of which two are in the double nucleus and the third is in an extended disc structure elongated SW-NE but separate from the dust lanes observed in the optical (see Figure 6.3).

The large luminosities of Arp 220 in the infrared and in CO (1–0) emission makes it an ideal target for an extragalactic molecular census. Low J lines of CS, HCN, and HCO$^+$ have been amply observed in this galaxy and they all imply

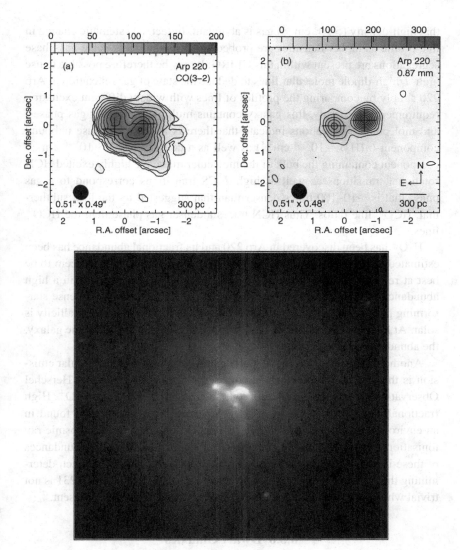

Figure 6.3. The ultraluminous merging galaxies Arp 220. Top: (a) Integrated intensity in CO(3–2) emission, from SMA observations. The plus signs mark the peaks of continuous emission at 345GHz, shown in (b). Reproduced by permission of the AAS from Sakamoto, K., Wang, J., Wiedner, M. C., Wang, Z., Petitpas, G. R., Ho, P. T. P., and Wilner, D. J. 2008. *Astrophysical Journal*, 684, 957. Bottom: Hubble image of Arp 220 (Credit: R. Thompson et al. NICMOS, HST, NASA).

that high-density ($>10^5$ cm^{-3}) gas is abundant. In fact, a systematic change in line profile as higher densities are probed suggests that large-scale gas-phase segregations are present within this ULIRG. It may be therefore possible to use high-J/high-dipole molecular lines to deduce the state of gas relaxation in Arp 220 simply by comparing the profiles of lines with widely different excitation requirements. Certainly, this galaxy contains multiple molecular gas phases, and molecular observations indicate that there is at least a diffuse unbound component ($n(H_2) \sim 10^{2-3}$ cm^{-3}) as well as a denser ($n(H_2) \sim 10^{5-6}$ cm^{-3}) component containing the bulk of the molecular mass: the highly excited HCN rotational transitions as well as high-J CS transitions correspond to a gas phase that is $\sim$10–100 times denser than that suggested by the low, subthermal, HCO$^+$ line ratios. Thus HCN traces a denser gas phase than the HCO$^+$ lines.

H$_3$O$^+$ has been discovered in Arp 220 and its fractional abundance has been estimated to be quite high (2–10×10^{-9}). Although X-ray models seem to be best at reproducing the observations of Arp 220, one can also obtain a high abundance of this ion at low (i.e., in a PDR) as well as high (i.e., dense star-forming gas) extinction as long as ζ is at least 10^{-13} s^{-1} and the metallicity is solar. At high extinction, which could represent the nuclear part of the galaxy, the abundance is even higher.

Another interesting object that has been recently studied in molecular emission is the ULIRG Mrk 231. A high-resolution spectrum from the Herschel Observatory reveals the presence of ions such as OH$^+$, CH$^+$, and H$_2$O$^+$. High fractional abundances (at least $\sim$$10^{-10}$) of these three ions can be found in an environment with low metallicity (0.1 times solar), fairly high cosmic ray ionisation rates ($\sim$$10^{-16}$ s^{-1}), and low visual extinction. The high abundances of these ions are often explained by involving XDR chemistry. In fact, determining the origin of molecular emission from ULIRGs such as Mrk 231 is not trivial when both sources of energy (cosmic rays and X-rays) are present.

6.3.6 Dwarf Galaxies

Despite the fact that many dwarf galaxies do form stars so that interstellar gas must be present, molecular detections in such objects are notoriously unsuccessful. It is yet not clear whether the lack of widespread CO implies a lack of molecular hydrogen or whether in these objects CO becomes a poor tracer of H$_2$. These objects are interesting because they are supposed to be crucial building blocks of much larger galaxies, yet there is mounting evidence that many dwarf galaxies in our local universe have stellar populations much younger than their larger cousins. A major source of this ambiguity may arise from the fact that accurate measurements of the interstellar properties of dwarf galaxies have been lacking for the most part. In recent years efforts to detect CO have been successful for a wide range of low-metallicity dwarf galaxies, and there seems to be a correlation between the presence of CO in a dwarf galaxy and its formation rate of high-mass stars. However, it is also found that non-detections are correlated with metal-poor low-mass galaxies, possibly implying that in this regime CO does not trace molecular hydrogen.

To date, no other molecules have been detected in dwarf galaxies and CO remains our best hope of tracing star-forming gas in these galaxies.

6.4 Star Formation and the Initial Mass Function

In Section 4.5 we discussed the various timescales that determine the formation of low-mass stars and showed in Figure 4.3 how these timescales vary with changes in the local parameters of density, metallicity, cosmic ray flux, and UV intensity. The figures clearly demonstrate that there are parameter combinations in which low-mass star formation is permitted (as in the Milky Way), whereas other combinations seem to preclude low-mass star formation. Assuming that the formation of high-mass stars is unrelated to the formation of low-mass stars (high-mass stars are usually considered to be formed in a sequential process triggered by major dynamical events such as nearby supernovae), then certain physical conditions may generate a conventional Initial Mass Function (like that in the Milky Way Galaxy; with large numbers of low-mass stars for each high-mass star) while other conditions may lead to a 'top heavy' Initial Mass Function (IMF), that is, having a relative lack of low-mass stars compared to the number of high-mass stars.

Conditions likely to suppress low-mass star formation – and lead to a top heavy IMF – include galaxies with intense UV fields and high cosmic ray fluxes in which the metallicity is near-solar. These conditions may be found in active galaxies at high redshift. Another galaxy type that may show a top heavy IMF are galaxies similar to the Milky Way but with subsolar metallicities. If these arguments are correct, then not only may it be possible to predict the form of the IMF in distant galaxies, but we may also be able to use molecular emissions to support that prediction. The CO rotational emission integrated antenna temperatures appear to depend significantly on the metallicity of the galaxy and on the activity of the galaxy (see Figure 6.4). In addition, emission from other molecular species (such as HCN, HNC, and CN) may usefully constrain the metallicity. Therefore, a coordinated approach may be able to determine roughly the IMF and the metallicity of high-redshift galaxies.

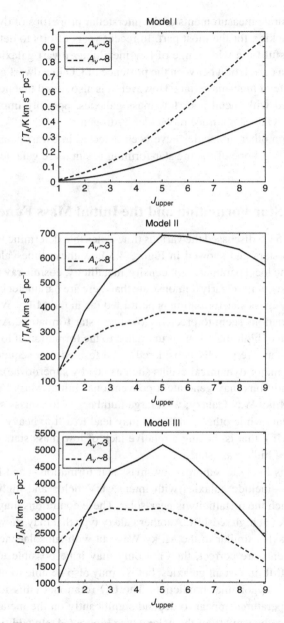

Figure 6.4. Theoretical velocity-integrated antenna temperatures for models of three high-redshift galaxies at $A_V = 3$ (solid lines) and 8 (dashed lines). Models I, II, and III have hydrogen number densities of 10^4, 10^5, and 10^5 cm^{-3}; metallicities of 0.05, 0.05, and 1.0 × solar; and cosmic ray ionisation rates 10^{-13}, 10^{-14}, and 10^{-14} s^{-1}, respectively. (Reproduced with permission from Banerji M., Viti, S., and Williams, D. A. 2009. *Astrophysical Journal*, 703, 2249.)

Figure 6.5. Image of part of the Perseus Cluster. The dominant galaxy in the cluster, NGC 1275, is on the left of the image. (Courtesy of ©: R. Jay Gabany).

6.5 Molecules in the Perseus Cluster of Galaxies

We have seen that interacting galaxies may give rise to starbursts, by abruptly feeding new interstellar matter into the star-forming process. Large clusters of interacting galaxies may have even more dramatic consequences: they may generate a new phenomenon not present in single or double galaxies.

The best studied galaxy cluster is the Perseus Cluster, centred on the galaxy NGC 1275. This cluster is, at about 80 Mpc, one of the closest galaxy clusters to Earth. The cluster generates powerful X-ray and radio emission as a result of material falling onto NGC 1275. The Perseus Cluster (see Figure 6.5) is part of a larger complex, the Pisces–Perseus Cluster; this larger cluster contains more than a thousand galaxies.

The central galaxy of the Perseus Cluster is surrounded by 'filaments' observed in optical emission lines, in infrared H_2 emission lines, and in CO millimetre-wave emission lines (see Figure 6.6). These filaments are huge: they are about 100 pc thick and extend for many tens of kpc. Their origin and excitation remain to be confirmed, but it is likely that they are set up as turbulent mixing layers between very fast diffuse outflows and entrained gas and that they are heated by dissipation or by cosmic rays.

However, the heating rates required to maintain the observed emissions from the filaments are very powerful compared to typical heating rates in the Milky

Figure 6.6. A Hubble image of NGC 1275 and its filaments. (Courtesy of NASA, ESA, and the Hubble Heritage (STScI/AURA)-ESA/Hubble Collaboration; A. Fabian.)

Way, and so molecular probes of the filaments could help to determine these rates and identify the heating source. Because CO emission has been detected, it seems natural to search for new probes in the form of other molecular emission lines. Models covering a very wide range of physical conditions in the filaments suggest that HCO^+, C_2H, and CN should be good probes, and recent detections of $CN(2-1)$, $HCO^+(3-2)$, and $C_2H(3-2)$ support the view that the main heating in the filaments could be supplied by cosmic rays, at a rate at least two orders of magnitude larger than in the Milky Way Galaxy.

Evidently, targeted chemical modelling and molecular line observations can be very useful in probing cloud conditions in exotic regions.

6.6 Conclusions

It is clear that molecules play a key role in the formation and evolution of galaxies. Recent years have witnessed a large number of extragalactic molecular surveys that have brought detailed information about the amount and distribution of the molecular gas in different types of galaxies. In the near future, especially with ALMA, the intermediate-to high-redshift Universe will be *revealed* in molecular emission at unprecedented spatial resolution. It is therefore likely that in the next decade or so we will be able to use molecular emissions from distant galaxies to explore the physical conditions in them and their likely evolutionary status, as we routinely now do for the Milky Way.

6.7 Further Reading

Astronomy & Astrophysics. 2010. Special Issue, 518. Herschel: The first science highlights.

Astronomy & Astrophysics. 2010. Special Issue, 521. Herschel/HIFI: First science highlights.

Fukui, Y., and Kawamura, A. 2010. Molecular clouds in nearby galaxies. *Annual Review of Astronomy and Astrophysics*, 48, 547, 2010.

Hartquist, T. W., and Williams, D. A., eds., 1998. *The Molecular Astrophysics of Stars and Galaxies*, New York: Oxford University Press.

Kennicut, R., and Evans, N. J. II. 2011. Star formation in the Milky Way and nearby galaxies. *Annual Review of Astronomy and Astrophysics*, 50, 119.

Omont, A. 2007. Molecules in galaxies. *Reports on Progress in Physics*, 70, 1099.

Solomon, P. M., and Vanden Bout, P. A. 2005. Molecular gas at high redshift. *Annual Review of Astronomy and Astrophysics*, 43, 677.

7

The Early Universe and the First Galaxies

We have seen in preceding chapters that molecular lines are excellent tracers of interstellar gas, of star-forming regions, and of the interactions of stars on their environments in the Milky Way Galaxy and in external galaxies. Observations of molecular emissions, supported by detailed modelling, allow a rather complete physical description to be made of the regions where these molecules are located, even when the galaxies are not spatially resolved. But what about pregalactic situations in the Universe? These include some of the most active areas of research in modern astronomy. Did molecules have a role to play in pregalactic astronomy, and if so could molecular emissions help to trace processes occurring very early in the evolution of the Universe? When did molecular processes begin to play an important role? What are the best tracers of the first galaxies in the Universe?

In this chapter we show that molecules were likely to be present from the era of recombination after the Big Bang and certainly played an important role in the formation of protogalaxies and of the first stars. Whether molecules generated detectable signatures of those very early events is problematic, at least with our present range of astronomical instrumentation. However, it seems likely that we shall soon find molecular signatures of the post-recombination era. Once the first stars appeared and seeded their environments with metallicity, the formation of the first galaxies was modulated by molecules and it should certainly be possible to trace their formation using molecular emissions.

7.1 The Pregalactic Era

7.1.1 Pregalactic Chemistry

In the conventional Λ-CDM picture, the Universe expanded and cooled from a hot dense state in which nucleosynthesis in the baryonic matter of subatomic particles produced a simple gas containing atomic nuclei of hydrogen, deuterium, helium 3 and 4, and lithium. Table 7.1 summarises the physical characteristics at the key stages in the history of the Universe: initially, all the gas was totally ionised and bathed in a black-body radiation field characterised by the same temperature as the gas. Hydrogen accounted for 75% of the baryonic mass, helium almost all of the remainder, and lithium a tiny trace. As the expansion continued, both the gas and the radiation cooled and eventually reached a temperature at which recombination enabled neutral atoms to survive. Recombination began first with helium ions to form neutral helium atoms (at a redshift, z, of around 2500) and later on with hydrogen ions (at $z \sim 1100$).

Once the bulk of the gas was neutral, then the temperature of the radiation and of the matter became essentially decoupled. The cosmic microwave background (CMB) radiation detected today is the relic of that radiation field from the recombination era when the temperatures of radiation and of gas were both about 3000 K at the end of the recombination era. The energy density in the radiation field was then diluted and transformed in the expanding Universe to become a black-body radiation field of temperature around 2.7 K at the present epoch.

The evolution of matter in the gas after decoupling was influenced by simple atomic and molecular processes in the expanding Universe, and radiation played a rapidly decreasing role. The energy density in the radiation declined very rapidly with decreasing z as $(1 + z)^4$ and the radiation temperature as $(1 + z)$, while the number density in the gas declined with decreasing z as $(1 + z)^3$. Therefore, as z decreased, the number of photons capable of ionising or photodissociating atoms or molecules declined catastrophically.

Recombination in the expanding Universe was not entirely complete, and a residual level of ionisation, with $n(e)/n(H) \sim 10^{-3}$ at the end of the recombination era-remained and continued to fall, being an order of magnitude smaller by $z \sim 100$. The fractional ionisation in the gas at that epoch was capable of initiating a simple chemistry. In the absence of dust grains, hydrogen molecules were formed in two sequences of reactions:

$$\mathrm{H} + \mathrm{H}^+ \rightarrow \mathrm{H}_2^+ + h\nu \tag{7.1}$$

$$\mathrm{H}_2^+ + \mathrm{H} \rightarrow \mathrm{H}_2 + \mathrm{H}^+ \tag{7.2}$$

and

$$\mathrm{H} + \mathrm{e} \rightarrow \mathrm{H}^- + h\nu \tag{7.3}$$

$$\mathrm{H}^- + \mathrm{H} \rightarrow \mathrm{H}_2 + \mathrm{e} \tag{7.4}$$

Table 7.1. Physical characteristics of the key stages in the history of the Universe

	Recombination era begins	Dark ages	Pop. III appear	Protogalaxies form	Pop. II appears	Present epoch
z	~1100	~1100–20	~20	~20–10	~10	0
t (yrs)	3.8×10^5	3.8×10^5–1.8×10^8	1.8×10^8–4.8×10^8	4.8×10^8	4.8×10^8–9.5×10^8	13.7×10^9
$\bar{n}_H$ (cm^{-3})	~6700	6700–0.046	0.046	0.046–6.78×10^{-3}	6.78×10^{-3}	5×10^{-6}
T_{CMB} (K)	~3000		~60	~60–30	~30	2.7
Metallicity (solar)	0	0	~10^{-4}	~10^{-4}–~10^{-3}	~10^{-3}	1

The redshift, z, the age of the Universe, the mean H-atom number density, the cosmic background radiation temperature, and the probable metallicity are given at various stages of evolution of the Universe.

Because H_2^+ is more strongly bound than H^- the first of these sequences was important at about $z \sim 400$ while the second dominated at $z \sim 100$, at which epoch the radiation field had moved to longer wavelengths and was less intense.

Helium may have also provided an additional route to H_2 through sequences of reactions such as:

$$He + H^+ \rightarrow HeH^+ + h\nu \qquad (7.5)$$

$$HeH^+ + H \rightarrow H_2^+ + He \qquad (7.6)$$

In fact, a surprisingly extensive chemical network was established from the three elements hydrogen, helium, and lithium, and their ions, with some species being chemically active in excited electronic or internal states. Deuterium, present at a level of about 10^{-5} of hydrogen, formed several species, including HD through reactions such as

$$H_2 + D^+ \rightarrow HD + H^+ \qquad (7.7)$$

Figure 7.1 shows the remarkable variety of atomic and molecular species that formed in the recombination era. Some of the basic reaction data still remain uncertain, but the fractional abundances of molecules and molecular ions computed in these and similar chemical models are very small indeed.

7.1.2 Potential Observational Consequences

The molecular abundances in the recombination era are so small (see Figure 7.1) that transitions in those species have not so far been directly detected through line emission. However, a great deal of attention has been given to the possible role of molecules in modifying the spectrum of the radiation field in the recombination era. Any such effects would appear in the present epoch as deviations of the observed CMB from a black-body form. These deviations would arise from optical depth effects created by line absorption, photoionisation, and photodissociation of atoms and molecules in the recombination era.

Optical depth effects on the radiation field would arise particularly from resonance line transitions of high dipole moment, such as transitions in HeH^+ and HD^+. Free–free transitions or photodetachment transitions in negative ions such as H^- and He^- have been explored, and photodissociation transitions of HeH^+ have been considered. Recent studies suggest that the H^- ion may have the greatest effect on the CMB profile, the distortion due to H^- being about one order of magnitude below the detectable limit of the Planck Space Observatory. Evidently, it is reasonable to hope that the H^- deviation, and possibly effects from other species, may become accessible in future CMB

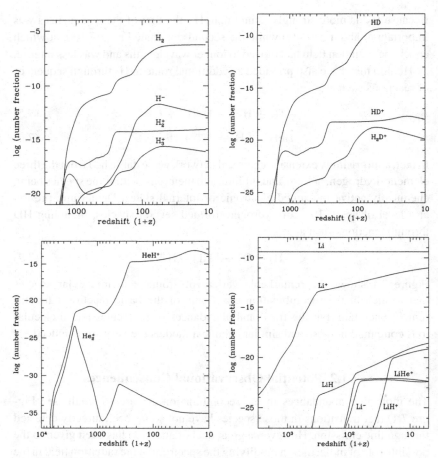

Figure 7.1. The evolution of chemical species formed in the recombination era, as a function of redshift. All species are initially atomic ions, and integration of the chemical and thermal equations is carried through the recombination era to the present epoch. (Private communication: D. Galli and F. Palla.)

observations, especially by the ALMA facility, which will be exceptionally sensitive in the millimetre waveband.

7.2 Formation of the First Stars

7.2.1 Chemistry in the Inhomogeneous Dark Ages

We have seen that in the near-homogeneous prestellar and pregalactic era the fractional abundance of molecular hydrogen was very low, at around 10^{-6}, and that other molecular species had very much lower abundances than H_2 (see Figure 7.1). However, as the redshift decreased, inhomogeneity in the Universe increased. According to the Λ-CDM model, small dark matter minihaloes were the earliest to form, and larger dark matter haloes formed later. The mainly atomic baryonic matter tended to fall into the gravitational wells of the minihaloes, and if the thermal pressure was not too great atomic matter began to accumulate in the haloes. Consequently, the larger number densities of baryonic matter present in the haloes increased the rate of formation of molecular hydrogen above that in the earlier near-homogeneous phase of the Universe, and much larger fractional abundances of H_2 were formed.

If the fraction of H_2 was high enough, then the rapid cooling ensured that the gas in a minihalo maintained a low enough temperature so that accumulation of baryonic matter could continue until the formation of the very first stars in the Universe, the Population III stars, occurred. Minihaloes at $z \sim 20$–30 are now considered to be the formation sites of Population III stars. On the other hand, if the cooling was too slow compared to the universal expansion timescale, then the temperature remained high and star formation was precluded. Therefore, through its cooling function, molecular hydrogen set a minimum mass for a minihalo to proceed to form stars. That is, larger dark matter minihaloes accumulated more atomic hydrogen that then created more molecular hydrogen; sufficiently effective cooling by molecular hydrogen enabled the formation of Population III stars to proceed.

Molecular hydrogen is an effective coolant down to temperatures $\sim$200 K until the number density rises to the critical value (see Section 2.3; for molecular hydrogen the critical number density is $n_H \sim 10^4$ cm^{-3}) above which collisions are dominant in relaxing the excited levels. For minihaloes with gas above this density, both cooling and mass accumulation were slower than in the lower density regime, but eventually reached the critical Bonnor–Ebert mass at which gravitational instability overcame thermal support. For $n_H \sim 10^4$ cm^{-3} and $T_K \sim 200$ K, the Bonnor–Ebert mass is around 10^3 $M_\odot$. For masses above this limit, the collapse continued, the density increased, and three-body reactions dominated in increasing the H_2 fraction. Ultimately, the cooling transitions of H_2 became optically thick and the cooling efficiency was reduced. The temperature rose as the collapse continued, H_2 was dissociated, and the temperature rise continued until thermal pressure caused the collapse to cease. This pressure-supported object is regarded as a Population III protostar.

Numerical simulations suggest that the protostar may have been surrounded by an accretion disk that may have grown sufficiently to become unstable. If so, then multiple protostars may have been formed from the disk material. In fact, molecular hydrogen may also have been a crucial factor in controlling the disk temperature; with H_2 cooling, the disk would have been at a lower temperature than without it, and therefore the disk would have been less able to transfer mass to the protostar. Thus, the disk mass would have grown into a phase of instability.

7.2.2 Potential Observational Consequences

Observations are not yet able to probe the era of $z \sim 20$ in which the formation of Population III stars occurred, but advances in technology may soon make that possible. At present, the deepest observations of the Universe using the Hubble Space Telescope identify galaxies at redshifts of $z \sim 7\text{--}8$, and the galaxy with the highest redshift currently known is at $z = 8.6$. Broad-band photometry can be used to identify absorption breaks due to neutral hydrogen and Lyman-α emission can also be a strong signature of emission from established galaxies. These features are probably shifted into the near-infrared. The Giant Magellan Telescope, the Thirty Metre Telescope, and the European Extremely Large Telescope are ground-based facilities that will very significantly enhance future observational possibilities. The James Webb Space Telescope, operating in the range 0.6–28 μm, should provide a unimpeded view of these atomic hydrogen features.

As described previously, the formation of Population III protostars should have been accompanied at all stages of evolution by copious H_2 emission in rotation-vibration bands. Of these, some of the redshifted ro-vibrational transitions may become accessible to the JWST. Pure rotational transitions may be redshifted into the submillimetre regime and could be a possible target for ALMA. Hydrogen atom emission at 21 cm should be redshifted into the metre regime and may become a target for observations using the Square Kilometre Array (SKA), which will initially, cover the frequency (wavelength) range 70 MHz (4 m)–10 GHz (3 cm). Therefore, there is a good prospect that this era of the formation of the first protostars (at $z \sim 20$) may soon become accessible to observation.

7.3 Formation of the First Galaxies

7.3.1 Chemistry after the Formation of Population III Stars

The conventional scenario for the formation of the first galaxies is summarised in Figure 7.2. We regard the formation of a galaxy as being characterised by the formation of the second-generation (Population II) stars. So, in this discussion, we are concerned with star formation that is assumed to have been similar in

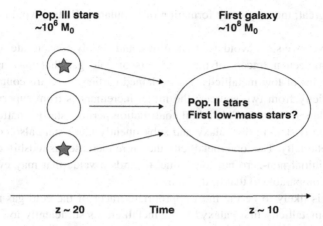

Figure 7.2. The first stars, Population III, were formed in dark matter haloes at an epoch around $z \sim 20$. The supernovae generated by the Population III stars in them may have caused a strong negative feedback on the haloes themselves, by injecting energy from the supernova explosions and from radiation, and the haloes may have cooled by atomic line emission. The assembly of the first galaxies from the haloes then provided the opportunity for the formation of the second generation of stars (Population II) in the interstellar medium that had been enriched in metallicity. (Re-drawn and modified from a figure by Bromm, V., and Yoshida, N. 2011. *Annual Reviews of Astronomy & Astrophysics*, 49, 373.)

nature to that occurring at the present time in massive molecular clouds in the Milky Way and in nearby galaxies, although the metallicity may have been very much lower.

For chemistry, the importance of the Population III stars that preceded the formation of the first galaxies is that they supplied the initial enhancement to the metallicity of their environments. Atoms of heavy elements such as oxygen, carbon, and nitrogen, and possibly dust grains incorporating some of those elements, were supplied to the gas for the first time, and distributed widely in the halo gas through turbulence and halo mergers.

The opportunity for a significantly enhanced chemical complexity in the Universe had arrived. With that complexity came a change in the star formation process, from the metal-free formation of Population III stars to the more famil-iar metal-rich process that generated early Population II stars, and continues to do so at the present epoch (see Chapter 5).

It is unclear at present whether dark matter haloes had already attained a critical (but low) metallicity, $\sim 10^{-6} - 10^{-4}$ of solar, enabling Population II stars to exist even before the formation of the first galaxies. Of course, if the thermal and dynamical feedback from the Population III stars to the haloes was too great, the very early formation of Population II stars would not have occurred.

However, we shall avoid such questions, and simply concentrate our attention on molecular tracers of the reservoirs of cold gas that may appear in first galaxies of low metallicity. As explained earlier, there are contributions to metallicity from two sources: the initial increment is from supernovae of the Population III stars. This initial contribution permits star formation of the Population II stars in the galaxy and subsequently these stars also contribute to the metallicity. The total metallicity therefore increases as redshift declines, from an initial non-zero but low value, towards a value that may eventually become comparable to that of the Sun.

What is likely to be the most appropriate tracer of the cold gas reservoir in a low-metallicity first galaxy? If the metallicity is sufficiently low in a gas that is mostly H_2, then H_3^+ can be surprisingly abundant and is proportional to the cosmic ray ionisation rate. As discussed earlier (see Section 5.1.2), cold molecular hydrogen is essentially undetectable, and so carbon monoxide has traditionally been the most useful tracer of cold gas in both nearby and high-redshift galaxies. A tight correlation has been observed by many astronomers to exist between the far infrared luminosity – regarded as a measure of the surface density of star formation – and the surface density of CO emission in its rotational lines. In the local Universe, this linear correlation extends over several orders of magnitude in far infrared luminosity, and encompasses a range of different environments from relatively low metallicity and largely atomic gas to relatively high metallicity and largely molecular gas.

It is particularly significant for our present discussion that this correlation has been confirmed for a sample of galaxies including those with red shifts up to $z \sim 6$, and also to be established in several different CO lines (see Figure 7.3).

Evidently the importance of CO as a tracer of galactic reservoirs of cold gas is confirmed in a variety of physical situations and of values of metallicity, and in galaxies with fairly high redshifts. Conversion factors (i.e., X-factors; see Section 5.1.2) have been computed for a range of physical conditions that may be relevant for high-redshift galaxies, including values of metallicities as low as 1% of solar, for several CO rotational transitions. The physical conditions adopted create kinetic temperatures of the interstellar gas in the range $\sim$10–100 K. Some of the results for very low metallicity cases are shown in Table 7.2; they differ significantly from values listed in Table 6.4 for solar metallicity.

Because X_{CO} is proportional to H_2 mass, it is important to use the correct conversion factor and to note the sensitivity to parameters. For example, if the (7–6) transition is being used, then the conversion factor is very sensitive to the ultraviolet field impinging on the cloud, whereas the X-ray intensity has a

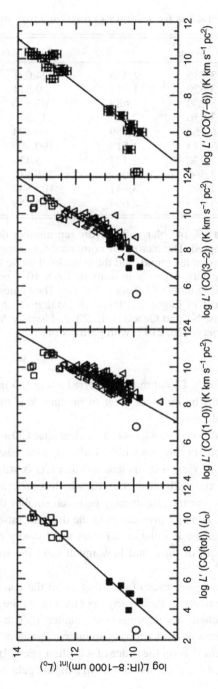

Figure 7.3. The relationships between integrated infrared luminosity and luminosity in CO (total line emission), (1–0), (3–2), and (7–6) transitions. Different symbols refer to data from different objects; open squares are data points for high-z sources. (Reproduced with permission, from Bayet, E., Gerin, M., Phillips, T. G., and Contursi, A. 2009. *Monthly Notices of the Royal Astronomical Society*, 399, 264.

Table 7.2. *Values of X—factors for several CO rotational transitions*

G_{UV}	F_X	$X_{CO(1-0)}$	$X_{CO(3-2)}$	$X_{CO(7-6)}$	T_K/K
1	0.01	3.585	2.335	24.091	39.37
1	0.1	3.620	1.817	9.311	51.24
1	1	6.260	3.688	63.556	107.78
10	0.01	3.856	2.393	27.359	38.93
10	1	6.468	3.759	66.494	104.35
100	0.01	3.673	3.513	190.302	38.17
100	1	13.730	2.477	5.428	98.49
1000	0.01	4.560	4.061	295.512	39.35
1000	0.1	2.592	3.841	102.544	19.58
1000	1	15.475	2.664	5.859	113.55

The metallicity is set to 1% of solar and a hydrogen number density in the cloud of 10^4 cm^{-3} is adopted. The factors are computed for a range of UV and X-ray intensities incident on the cloud, and the resulting kinetic temperatures are shown. The UV intensity, G_{UV}, is in units of 1.6×10^{-3} erg cm^{-2} s^{-1}, and the X-ray intensity is in units of 1.0 erg s^{-1} cm^{-2}. The values in this table are extracted from Table 1 of Lagos, C., Bayet, E., Baugh, C. M., Lacey, C. G., Bell, T. A., Fanidakis, N., and Geach, J. E. 2012. *Monthly Notices of the Royal Astronomical Society*, 426, 2142, with permission.

significant but smaller effect. Cloud masses inferred using CO integrated line intensities may be in error by up to an order of magnitude if these radiation parameters are not well determined.

Thus, from an observational standpoint, it is clear that CO is an effective tracer of cold gas reservoirs in galaxies with redshifts up to at least 6. However, from a theoretical point of view it is unclear whether CO is still an effective tracer of cold gas in galaxies with redshifts ~ 10, which is the epoch of galaxy formation. At that epoch, the metallicity may be much smaller than the $\sim 1\%$ of solar assumed earlier. A major problem is in the dust:gas ratio. How much of the metallicity supplied by Population III stars and by very early Population II stars is in the form of dust, and how much is in the form of heavy atoms?

There are (at least) two properties of dust that affect the considerations of 'normal' interstellar chemistry at these early epochs. One is the formation of H_2 on which almost all chemistry depends (see Chapter 3). The timescale for formation of H_2 in reactions on dust grains in an interstellar cloud of the Milky Way with 10^3 H atoms cm^{-3} is on the order of a million years. If the dust:gas ratio scales with metallicity, and the metallicity in a high-z galaxy is, say, 10^{-3}

of solar, and assuming that the dust has the same efficiency to form H_2 as dust in the Milky Way, then the H_2 formation timescale by surface reaction was on the order of a billion years. This may be longer than a cloud lifetime in an early-time galaxy. Depending on the fractional ionisation in the gas, the processes initiated by electrons and by protons in radiative association with H atoms (see Section 7.1.1) may have been faster routes to forming H_2 in a first galaxy, and may therefore have enabled a conventional interstellar chemistry to appear.

Assuming that H_2 could be made on a reasonable timescale, then the formation of CO in a first galaxy occurred in a PDR (see Figure 4.1; Section 4.1.1). For average conditions in the interstellar medium of the Milky Way Galaxy, the conversion of carbon into CO occurs at a depth of about 2–3 visual magnitudes into an interstellar cloud, which corresponds to a distance that is typically on the order of a parsec. If the dust:gas ratio in a first galaxy was 10^{-3} of that in the Milky Way with dust of the same optical properties, the corresponding depth would have been on the order of a kiloparsec. It is unclear whether that is an acceptable scale size for clouds in a first galaxy. The role of XDRs in the first galaxies in modifying the abundaces in the PDR also needs to be evaluated.

Many questions concerning chemistry in the first galaxies remain to be resolved. Although detailed studies of chemistry at very low metallicities are awaited, the best practice for observational strategy seems to be to continue to regard CO as the most reliable tracer. Of course, hydrogen atom radio recombination lines may also be detectable in the first galaxies with very low metallicity, but these will not constrain the cold gas reservoir. They may, however, help to constrain the UV and X-ray field intensities.

7.3.2 Potential Observational Consequences

At redshifts $z \sim 10$, the H atom 21-cm line is shifted into the metre waveband and therefore becomes an important target for Square Kilometre Array (SKA) observations. Thus, the atomic component of the interstellar gas should become well determined. Of greater interest for the problem of star formation is the determination of the molecular component of the interstellar gas. The lowest energy CO transition (1–0) shifts from 2.6 mm to $\sim$2.6 cm, which may become accessible to the SKA, if not initially then eventually in Phase 3 of that project. Also, H-atom radio recombination lines should be accessible to SKA. In the case of H_2-rich gas of very low metallicity, H_3^+ is a potential tracer. It has neither an electronic spectrum (no electronically excited bound states) nor a pure rotational spectrum (no dipole moment) but it does have a ro-vibrational spectrum (not redshifted) near 3 μm. Thus, these lines could be accessible to the JWST for redshifts less than about 10.

7.4 Further Reading

Bromm, V., and Yoshida, N. 2011. The first galaxies. *Annual Review of Astronomy and Astrophysics*, 49, 373.

Clark, P. C., Glover, S. C. O., Smith, R. J., Greif, T. H., Klessen, R. S., and Bromm, V. 2011. The formation and fragmentation of disks around primordial protostars. *Science*, 331, 1040.

Galli, D., and Palla, F. 2013. The dawn of chemistry. *Annual Review of Astronomy and Astrophysics*, 51, in press.

Glover, S. C. O. 2011. *The Chemistry of the early Universe in the Molecular Universe*, edited by J. Cernicharo and R. Bachiller, p. 313. IAU Symposium 280. Cambridge: Cambridge University Press.

Lehnert, M. D., Nesvadba, N. P. H., Cuby, J. G., Swinbank, A. M., Morris, S., et al. 2010. Spectroscopic confirmation of a galaxy at redshift 8.6. *Nature*, 467, 940.

Loeb, A., and Barkana, R. 2001. The reionization of the Universe by the first stars and quasars. *Annual Review of Astronomy and Astrophysics*, 39, 403.

Schleicher, D. R. G., Galli, D., Palla, F., Camenzind, M., Klessen, R. S., Bartelmann, M., and Glover, S. C. O. 2008. Effects of primordial chemistry on the cosmic microwave background. *Astronomy & Astrophysics*, 490, 521.

8

Recipes for Molecular Submillimetre Astronomy

It is clear that molecular line emissions can yield important information about the physical conditions of the gas and dust in our own as well as in external galaxies. What is not so clear is how to transform observational results into such physically meaningful information. This chapter aims at providing some simple recipes to aid the observer in achieving this goal. In this chapter we describe the relationships between the observable quantities in submillimetre molecular astronomy and the physical information that the observer would like to obtain. Inevitably, several approximations are made in order to derive such information, depending on the spectral and spatial resolution available, as well as the number of observable molecular transitions of any one species.

We start by relating what we measure with submillimetre and radio telescopes, that is, the antenna temperature, T_a, to the fundamental molecular constants and the relevant astronomical parameters. What we would like to know are the column densities of the observed species and the gas temperatures and number densities of the local gas. We shall describe first how to obtain these quantities if the gas is in Local Thermal Equilibrium (LTE), that is, where the level populations are dominated by collisions in a gas at a uniform temperature. The methods of obtaining this information depend on the types of molecule involved. We conclude this chapter by discussing the conversion of column densities into fractional abundances compared to the total hydrogen abundance, quantities that are directly comparable to the predictions of chemical models. In Chapter 9 we describe some approximations routinely used when LTE does not apply.

In this chapter we make use, without derivation, of some standard equations of molecular physics; the derivations of these equations can be found in standard texts (see Further Reading at the end of the chapter).

8.1 The Antenna Temperature

In submillimetre astronomy, the observer usually obtains an antenna temperature, T_a, produced at a frequency ν by a source having an optical depth, τ:

$$\tau = \frac{h}{\Delta \nu} N_u B_{ul} (e^{h\nu/kT} - 1) \tag{8.1}$$

where N_u is the column density of the upper state (and usually this is the information the observer requires) and $\Delta \nu$ is the full width at half-maximum line width in units of velocity. This optical depth, τ, is computed by considering the probability that the emitted photon may escape from the relevant region, allowing for absorption at a rate $B_{ul}\rho$, where ρ is the energy density, and spontaneous emission at a rate A_{ul} along the path, and noting that:

$$A_{ul}/B_{ul} = 8\pi h\nu^3/c^3 \tag{8.2}$$

where A_{ul} and B_{ul} are the Einstein coefficients for transitions between upper level u and lower level l (see Chapter 2).

By considering the relationship between brightness, B_ν, and optical depth:

$$B_\nu = \frac{h\nu/k}{e^{h\nu/kT} - 1} \left(\frac{1 - e^{-\tau}}{\tau} \right) \tau \tag{8.3}$$

(where k is the Boltzmann constant) one can then express the antenna temperature as:

$$T_a = \frac{hc^3 N_u A_{ul}}{8\pi k\nu^2 \Delta \nu} \left(\frac{\Delta \Omega_s}{\Delta \Omega_a} \right) \left(\frac{1 - e^{-\tau}}{\tau} \right) \tag{8.4}$$

where $\Delta \Omega_s$ and $\Delta \Omega_a$ are the source and antenna solid angles. Note that $T_a \Delta \nu$ is in fact simply the integrated line intensity over frequency (expressed as velocity here, and a measurable quantity):

$$W = \int T_a \, dv = T_a \Delta v \tag{8.5}$$

In principle, N_u can be obtained by inverting Equation 8.4, in terms of the integrated line intensity, to obtain:

$$N_u = \frac{8\pi k\nu^2 W}{hc^3 A_{ul}} \left(\frac{\Delta \Omega_a}{\Delta \Omega_s} \right) \left(\frac{\tau}{1 - e^{-\tau}} \right) \tag{8.6}$$

It is worth noting that the expression above can be greatly simplified if the following two assumptions are satisfied: (1) the source fills the beam so that the term involving the solid angles may be ignored, and (2) the emission in the transition under consideration is optically thin so that τ is small enough that the final bracket is close to unity. Then N_u is directly proportional to the integrated intensity. If instead the source is not resolved then the correct upper level column density will be greater by the factor $\Delta\Omega_a/\Delta\Omega_s$.

Of course, Equation 8.6 gives us only the column density of a species in the upper state of the transition, whereas we are usually interested in the *total* column density of the observed species. Several approximations are often made to derive, from the antenna temperature of one transition and the level column density, the total column density of a species. The first point one needs to consider is whether the system is in LTE, that is, whether we can assume that all the thermodynamical properties have thermodynamic equilibrium values at the *local* values of temperature and pressures; we discuss the LTE case in the next section. In most interstellar environments, the hydrogen density is insufficient to thermalise some or all the transitions so that the populations of the energy levels of a certain species cannot be described by LTE and in this case radiative excitation followed by spontaneous emission may dominate over collisions (see Chapter 2). We will cover the non-LTE case in Section 8.3 and in Chapter 9.

8.2 Local Thermodynamic Equilibrium

In an environment that is in LTE the total column densities of the observed species as well as the kinetic temperature of the gas can be obtained from observation of a single transition (note, however, that other quantities, such as density and velocity structure, can still only be obtained with observations of multiple transitions). In LTE at temperature T_k the column density of each upper level u is related to the total column density via the following equation:

$$N_u = \frac{N}{Z} g_u e^{-E_u/kT_k} \tag{8.7}$$

where N is the total column density of the species, Z is the partition function, g_u is the statistical weight of the level u, and E_u is its energy above the ground state. Hence, in LTE, it is straightforward to obtain the total molecular column density from the column density of any transition once the temperature is known, as we show in Section 8.2.3 in two examples for selected species.

If more than one transition is observed, we can also use Equation 8.7 to construct a so-called *rotational diagram*: this relates the column density per statistical weight of a number of molecular energy levels to their energy above the ground state. The diagram is a plot of the natural logarithm of N_u/g_u versus E_u/k and, for Equation 8.7, is a straight line with a negative slope of $1/T_k$ (see Figure 8.1, which shows two components of different temperatures on this line of sight). The temperature inferred from such a diagram is often referred

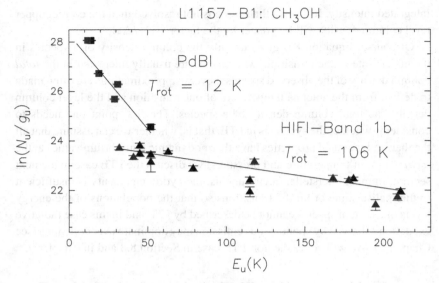

Figure 8.1. Rotation diagrams for the CH3OH transitions measured from space (triangles) and from ground (squares). The parameters N_u, g_u, and E_u are, respectively, the column density, the degeneracy, and the energy (with respect to the ground state of each symmetry) of the upper level. (Reproduced with permission from Codella, C., Lefloch, B., Ceccarelli, C., Cernicharo, J., et al. 2010. *Astronomy & Astrophysics*, 518, L112.) Copyright ESO.

to as the rotational temperature and it is expected to be equal to the kinetic temperature if all the levels are thermalised.

Such diagrams can be a useful tool to assess whether the level populations can be described in LTE and to determine the temperature that describes best the population distribution in the event that LTE applies. The diagrams can also be used to assess whether the emission is optically thin or thick: if we rewrite Equation 8.7 as:

$$\ln \frac{N_u}{g_u} = \ln N - \ln Z - \frac{E_u}{kT_k} \tag{8.8}$$

we can include the possibility of finite optical depth by considering an optical correction factor, $C_\tau = \tau/(1 - e^{-\tau})$ (from Equation 8.3):

$$\ln \frac{N_u^{\text{thin}}}{g_u} + \ln C_\tau = \ln N - \ln Z - \frac{E_u}{kT_k} \tag{8.9}$$

In the optically thin case the optical depth is very small, so that C_τ is close to unity. But in the optically thick case, the correction factor is non-zero, and Equation 8.8 will underestimate N.

So, in summary, an optically thin transition produces an antenna temperature that is proportional to the column density in the upper level of the transition being observed, and if all the transitions are thermalized and we know the kinetic temperature, we can convert the single measured column density into the total column density of the species in question. However, if the emission is not optically thin the finite opacity produces an underestimate of the upper level column density of the observed transition as well as of the rotational temperature.

The problem, of course, is that the observer is usually not aware whether the lines observed appear optically thin because the opacity is truly small or because the emission is beam diluted.

The effects of optical depth are different depending on whether the molecule is linear or not. In the following subsections we briefly discuss linear and nonlinear molecules.

8.2.1 LTE: Linear Molecules

As we saw in Chapter 2, for a linear molecule we can label each state by its rotational quantum number, J, its energy, $E_J \simeq BJ(J+1)$ (where B is the rotational constant, not to be confused with B_{ul} the Einstein B-coefficient), and its statistical weight, $g_J = 2J + 1$. Then the frequency of the transition $J \to J - 1$ is given by $\nu_{J,J-1} = 2BJ$, and the corresponding Einstein A-coefficient is given by:

$$A_{J,J-1} = \frac{64\pi^4 \nu^3 \mu^2}{3hc^3} \frac{J}{2J+1} \quad (8.10)$$

(recall from Section 2.1.2 that μ is the permanent electric dipole moment). The Einstein B-coefficient is therefore:

$$B_{J,J-1} = \frac{8\pi^3 \mu^2}{3h^2} \frac{J}{2J+1} \quad (8.11)$$

Using Equations 8.1, 8.7, and 8.11, we can now then write the optical depth of this transition as:

$$\tau_{J,J-1} = \frac{8\pi^3 \mu^2}{3h} \frac{N}{\Delta v} \frac{1}{Z} J e^{-(B/kT)J(J+1)} \left(e^{2(B/kT)J} - 1 \right) \quad (8.12)$$

For linear molecules in LTE, every term on the right-hand side of Equation 8.12 can be determined for an observed transition. Hence, the optical depth can be found.

8.2.2 LTE: Nonlinear Molecules

For nonlinear molecules the energy level structure can be quite complex, as we mentioned in Chapter 2. Often there are several transitions from a given state, and the dependence of transition frequencies on the energy of the upper level is not straightforward. Determining the optical depth for nonlinear molecules is therefore not trivial and has to be dealt with case by case. Calculations of the optical depth of each transition are affected by the fact that nonlinear molecules have (1) a complex energy level structure (e.g., several transitions from a given state may be allowed); (2) a complex dependence of transition frequencies on the energy upper state; (3) a dependence on a larger range of H_2 densities for thermalisation, and (4) a large range of absorption coefficients (i.e., the optical depths vary considerably for a given molecular abundance). In general, the effect of uncorrected optical depths cannot be ignored for nonlinear molecules. In the following subsections we list LTE formulae for two commonly observed interstellar molecules.

8.2.3 LTE Formulae for Two Commonly Observed Interstellar Molecules

The preceding brief description on the relationship between the spectroscopic characteristics of each species, observed quantities, and the physical conditions of the cloud will provide readers with some general formulae as a starting point for any molecule. However, it is clear from our discussion that depending on the geometry of the emitting region as well as the transition and species observed, different formulae will apply. Moreover, the equation for the column density in one transition derived previously can be written in different formalisms depending on which transition is observed, the type of region that is emitting it, the spatial resolution of the instrument, and whether all the relevant spectroscopic quantities for the transition are known. Because some molecules are observed more frequently than others, in this section we provide examples of practical formulae, without derivation, for two commonly observed molecules.

In Equation 8.7, once we know the column density in a particular state, we need to determine the partition function, whose elements are calculated differently depending on the molecule. We therefore start by providing expressions for the partition functions of different types of molecules and then provide readers with a form of Equation 8.6 for two molecular species.

In the following examples, we always assume that the molecular emission is optically thin.

Example of a column density calculation for NH$_3$

Ammonia is a symmetric top molecule, and two terms need to be considered in its partition function: the distribution across the metastable rotational levels ($J = K$) and the distribution between the two levels of the inversion transition within each rotational state. The partition function for the metastable states can be expressed as:

$$Z_J = (2J + 1)\exp\left\{-\frac{h}{kT}[BJ(J+1) + (C-B)J^2]\right\} \qquad (8.13)$$

where B and C are the rotational constants for ammonia (298 117 and 186 726 MHz, respectively). However, note that because *ortho-* and *para-*states are not expected to exchange, one has to write two separate partition functions for each state.

By also including the partition function elements of the inversion transition (for which the upper and lower states have equal statistical weight), the total partition function for, say, *para-*NH$_3$ is:

$$Z_{p-NH_3} = \sum_J \left[1 + \exp\left(-\frac{T_{ul}(J)}{T}\right)\right](2J+1)$$

$$\times \exp\left\{-\frac{h}{kT}[BJ(J+1) + (C-B)J^2]\right\} \qquad (8.14)$$

where $T_{ul} \equiv (h\nu/k)$ and for all the inversion transitions is ~ 1 K. Let us now take as an example the observed transition NH$_3$ (1,1); then its column density is:

$$N = \left[\frac{4(2\pi)^{3/2}}{c^3}\frac{g_l}{g_u}\frac{\nu^3}{A_{ul}}\right]\left[1 - \exp\left(-\frac{T_{ul}}{T_{ex}}\right)\right]^{-1}\tau\sigma \qquad (8.15)$$

where τ (the optical depth of the line) and σ (the observed line width) are observables and lead to an estimation of T_{ex}; hence by coupling all the constants for this transition (in CGS units), the expression can be simplified as:

$$N_{1,1} \approx 1.85 \times 10^{14}\left[1 - \exp\left(-\frac{1}{T_{ex}}\right)\right]^{-1}\tau\sigma \qquad (8.16)$$

By plugging the value obtained from Equation 8.16 into Equation 8.7, together with the value from the partition function, one can then find the total column density of ammonia in LTE.

Example of a column density calculation for N_2H^+

N_2H^+ is a linear rotator; hence the partition function is given by:

$$Z_J = \sum_J (2J + 1) \exp \left\{ -\frac{h}{kT} [BJ(J + 1)] \right\} \qquad (8.17)$$

where the rotational constant B is 46586.867 MHz. Let us assume we have observed the hyperfine structure of the N_2H^+ (1–0) emission; then the column density in this transition can be written as:

$$N = \frac{8\pi W}{\lambda^3 A_{ul}} \frac{g_l}{g_u} \frac{1}{T(R)_{ex} - T(R)_{bg}} \frac{1}{1 - \exp(-h\nu/kT_{ex})} \qquad (8.18)$$

where W is, as before, the integrated intensity of the line and $T(R)_{ex}$ and $T(R)_{bg}$ are the Rayleigh–Jeans excitation temperature and the temperature of a background source (i.e., the cosmic microwave background radiation).

Of course, for the hyperfine structure of N_2H^+, it is the calculation of the integrated line intensity that can be tricky owing to the presence of several lines within a small frequency range.

8.2.4 Practical Cases: Where LTE Applies

As mentioned earlier, LTE applies when collision transition rates dominate over radiative transitions. In reality, LTE is often never fully realized in the low densities of the interstellar medium. For any given transition, LTE is approached if the density exceeds a critical value, n_{crit}, given by the ratio of the Einstein A-coefficient over the collisional rate (see Chapter 2). The A-coefficient scales as $\mu^2 \nu^3$ where μ is, as before, the dipole moment. In practice, the decay rate is lowered by line trapping for optically thick lines so that the critical density is decreased. Typical values of n_{crit} thus range from $\sim 2 \times 10^3$ cm^{-3} for CO(1–0), because of the low value of the CO dipole, to $\sim 10^5$–10^6 cm^{-3} for the first transitions of molecules with large dipoles, such as HCN, CS, or HCO$^+$ (see Chapter 10 for more examples). However, this 'rule of the thumb' assumes a homogeneous emission region for each of the transitions observed. In reality, of course, the interstellar medium may be clumpy down to scales of hundredths of a parsec, as discussed in Chapter 5.

This means that although gas from within a beam of one transition may be in LTE, another transition of the same or a different species within the same beam may not be in LTE. Hence, we need to differentiate further and discuss resolved versus unresolved emission. If the medium is homogeneous then it is relatively straightforward to determine whether the observed transition is in LTE; however, a truly homogeneous medium implies very small spatial scales. So, in reality, it is only when the emission is resolved and we can spatially disentangle the different gas components that we can be certain that LTE conditions apply. In most cases, and certainly for unresolved sources (especially distant galaxies) the beam will encompass a range of densities and temperatures, and LTE calculations will necessarily give only crude estimates of the densities and temperatures of the gas, and the column density of the species observed. Densities and temperatures derived from LTE calculations should always be treated as estimates rather than exact values.

8.3 Non-LTE

If the medium is not thermalised, that is, if the hydrogen density is not sufficient to thermalise some or all of the transitions, a different temperature may characterise the population of each level relative to that of the ground state or relative to that of any other level. Recall from Chapter 2 that the excitation temperature T_{ex}, is defined by the relative populations or column densities of any two levels i and j of statistical weights g_i and g_j and energies E_i and E_j relative to an arbitrary common reference through the Boltzmann equation:

$$\frac{N_j}{N_i} = \frac{g_j}{g_i} e^{[-(E_j - E_i)/kT_{ex}]} \tag{8.19}$$

In non-LTE a region of space can be affected by the radiation field and the total column density can no longer be obtained by using Equation 8.7; radiative transfer has, then, to be locally computed. In Chapter 9 we describe the radiative transfer problem and discuss in detail an approximation for the radiative transfer routinely used in interstellar studies to derive the level populations.

8.3.1 Derivation of Fractional Abundances from Column Densities

So far in this chapter, we have been preoccupied with deriving column densities and their relationships with the gas density and temperature. However, observers often prefer to determine fractional abundances with respect to molecular hydrogen. In many cases, there is an advantage in obtaining fractional abundances: they are independent of geometry in that they give the *total* abundance (with respect to some form of hydrogen) of a species in the (assumed homogeneous) emitting region.

There are several methods used by astronomers to convert molecular column densities to fractional abundances, depending on the optical thickness of the line(s) available, and on the extra information the observer may have.

Obviously, if the column density of hydrogen within the same emitting region is known then the conversion is straightforward. Often, however, we do not have a direct measure of H_2 column density. Dust observations can provide us with measures of the optical depth, or visual extinction, of the region observed, and from them we derive the H_2 column density by making the usual assumption (for the Milky Way) that in 1 magnitude of material we have $\sim 1.6 \times 10^{21}$ cm^{-2} of hydrogen nuclei. However, if, as it is often the case, the beams from the dust observations differ from those of the molecular observations then there will be an error associated with the H_2 density, and hence with the mass of the region. Although this error is proportional to the beam dilution effect if the region is relatively homogeneous, it is not easy to estimate if clumpiness is present.

In many cases we do not have dust observations of the region in question, but we do have CO observations. In such cases the abundance of our molecule of interest can be derived as a fraction of CO; then a standard CO/H_2 ratio is assumed (via the X factor; see Section 5.1.2). There are at least two problems with this approach:

1. Unless the beam size of the observations of the observed species and of CO is the same, the fractional abundance estimate suffers from beam dilution effects. In fact, even if the beam size is the same but the observed source is smaller than the beam size, CO has a fairly low critical density with respect to most species. This is particularly important for large molecules, which tend to be tracing compact sources and therefore will be tracing a different gas component than is traced by CO.
2. The CO/H_2 ratio varies from galaxy to galaxy (see Table 6.4). Hence, determination of the H_2 density from CO observations may not be accurate due to the variation in this ratio. Theoretical studies show that for the CO(1–0) transition, the computed conversion factor may depart significantly from the canonical value for the Milky Way. Table 6.4 shows that the conversion factor for CO(1–0) ranges from about one order of magnitude above the Milky Way value to one order below, depending on the physical conditions in the galaxy observed. It also seems that for galaxies different from our own, the CO(1–0) line may not be the best tracer for H_2, and that higher J transitions of CO or even ionized carbon may be a better tracer.

In Chapter 9 we explore the reverse approach whereby theoretical column densities (derived from theoretical computations of fractional abundances) can be directly compared to observed column densities. This approach may be less prone to the large uncertainties listed here.

8.4 Further Reading

Draine, B. T. *Physics of the Interstellar and Intergalactic Medium*. Princeton Series in Astrophysics. Princeton, NJ: Princeton University Press.

Goldsmith, P. F., and Langer, W. D. 1999. Population diagram analysis of molecular line emission. *Astrophysical Journal*, 517, 209.

Tennyson, J. 2005. *Astronomical Spectroscopy: An Introduction to the Atomic and Molecular Physics of Astronomical Spectra*. London: Imperial College Press.

9

Chemical and Radiative Transfer Models

In Chapter 8 we covered the basic formulae and recipes that astronomers use to derive physical quantities from molecular observations. These simple LTE analyses provide observers with rough estimates of the density and temperature of the gas at equilibrium. However, molecular observations can also provide much further insight into the physical conditions and the history and dynamics of the gas if interpreted with the right tools. In this chapter we describe the chemical and radiative transfer models that have been developed over many years and we show how a careful use of such tools makes molecules into powerful diagnostics of the evolution and distribution of molecular gas in the interstellar medium. It is now possible for the *observer* to use well-established modelling codes to exploit the information contained in the observational data and to determine a rather complete description of the observed interstellar material. This chapter discusses the inputs required and the outputs expected from such models.

9.1 Chemical Modelling

Owing to the large range of densities and temperatures present in the interstellar medium, significant changes in the energetics and dynamics of the gas can occur, leading to large variations in the chemical abundances. For decades now, chemical simulations (based on the processes described in Chapter 3) have provided astrochemists with predictions of molecular abundances as a function of the physical conditions. However, the interpretation of chemical models is not a trivial task and demands a detailed knowledge of the way the chemical model is developed. There are at least four classes of chemical models: (1) the steady-state, single-point chemical models, used in the past mainly to predict equilibrium abundances in homogeneous media; (2) the

152

steady-state, depth-dependent models, more commonly called PDR models, used to predict abundances in photon-dominated regions; (3) time-dependent single-point models, commonly used to determine the abundances in dense cores; and finally (4) time-dependent depth-dependent models. This last type is a more recent addition to the suite of models. These models can be very useful in the determination of abundances in environments where not only the chemistry changes with time but also the dynamics of the region is such that local variations in the physical conditions lead to substantial variation in the chemistry as a function of position (e.g., shocked regions). In addition, each type of model may comprise a different type of chemistry, from purely gas-phase chemistry to very complex gas–grain chemistry. Finally, some PDR models have now been developed in 2D and 3D in order to cope with more complex geometry, for example, asymmetric planetary nebulae or gas around HII regions.

All classes of models essentially involve the computation of the solution of rate equations from astrochemical kinetics comprising a system of ordinary differential equations (ODEs) representing formation and loss terms for each species. The more complex the chemistry included, the more CPU-demanding is the solution of such system, as often the ODEs are 'stiff' owing to widely differing chemical and physical characteristic timescales as well as the presence of one or more fast processes in time, such as freeze-out, for example.

Chemical models can also be divided into two types: pure gas-phase models and gas–grain models. This latter category is discussed in Section 9.1.4.

9.1.1 What Goes into the Chemical Models?

In this section we briefly describe the physical and chemical input parameters within each class of model to which the user must assign numerical values. Varying these parameters allows users to investigate large-parameter spaces covering interstellar conditions that may be suitable for a variety of objects, from diffuse clouds to high-redshift galaxies. These parameters were introduced in Chapter 4.

Initial elemental abundances: Some models may start their chemical computations from a predefined gas and ice mantle composition, but most assume a predominantly atomic initial gas composition. Initial elemental abundances of carbon, oxygen, helium, nitrogen, sulfur, and metals need therefore to be set. These usually are set to solar or to scale with solar metallicity. However, in some situations the relative abundances of elements to each other are far from solar values. For example, the observed C/O ratio can vary significantly, or in the early Universe the nitrogen abundance may be suppressed (see Chapter 7).

Geometry: For PDR models as well as the chemical time- and depth-dependent models, the size (for 1D models) and shape (for 3D models) need to be specified; these specifications will affect the visual extinction of the simulation that in turn affects the local intensity of the electromagnetic radiation field and hence the photochemistry.

External cosmic ray ionisation rate (ζ) and radiation field strength (χ): In most reaction networks the rate of cosmic ray ionisation hydrogen is set to a typical value for the Milky Way Galaxy $\sim 10^{-17}$ s^{-1}. This rate can be enhanced or reduced in all chemical models. Similarly, in most networks the photoreaction rates are estimated at the standard averaged 'Draine' or 'Habing' interstellar radiation field energy density for the Milky Way (equivalent to 8.9×10^{-14} erg cm^{-3} and 5.3×10^{-14} erg cm^{-3}, respectively). In every model type a scaling factor can be applied to enhance or reduce the photoreaction rates.

Gas number density: For steady-state, single-point models the user will fix the density of the gas before the computation is started; for time-dependent and multipoint calculations often only the initial (and final, in the case of time-dependent) density is fixed; different density evolution functions can then be included in the model, depending on how the density is changing with time and/or depth (e.g., in a simulation of free-fall collapse).

Gas and dust temperature: Most PDR codes compute the thermal balance self-consistently by solving the level populations for the main coolants whose abundances have been determined in the model. Time/depth dependent chemical models, on the other hand, tend to have the temperature as a fixed parameter.

Dust grain properties: All chemical models include molecular hydrogen formation on grains; some also include other surface reactions and dust scattering. Hence, important parameters that need to be set by the user are the grain size or size distribution, and the grain scattering properties (e.g., albedo and phase function). A commonly used dust grain size distribution is the Mathis–Rumpl–Nordsieck (MRN) distribution, which makes a fit to the average interstellar extinction curve on the assumption that that dust consists of two materials, amorphous silicate and graphite. The extinction is best fitted on the assumption that all grains have a size distribution with $n(a)da \sim a^{-3.5}da$ where a is the radius of a dust grain. The albedo in the UV/optical regime is often assumed to be between 0.5 and 1. Dust models now generally go beyond the MRN model and include a PAH component. See Further Reading at the end of the chapter for descriptions of more recently computed dust size distributions and their optics.

Freeze-out: For models in which solid-state species are included, when simulating the cold phase of the interstellar medium, atoms and molecules from the gas will freeze on to the dust grains. In the absence of desorption, the rate per unit volume at which a species depletes on the grain can be approximated by the following proportionality:

$$\frac{dn(i)}{dt} \propto d_g a^2 n_H S_i T^{1/2} m_i^{1/2} \tag{9.1}$$

where d_g is the ratio of the number densities of grains to hydrogen nuclei, a is the grain radius, T is the temperature of the dust, and m_i and n_i are the mass and the number density of species i, respectively. S_i is the sticking coefficient, a number in the range of 0 to 1, and is often treated as a free parameter in chemical models as little experimental guidance is available for most species. For weakly bound species, such as CO and N_2, experiments suggest, however, that S_i is close to unity. The effects of freeze-out were discussed in Section 5.1.3.

Desorption processes: Thermal and nonthermal desorption processes induced by H_2 formation, and by cosmic rays and photons, are included in some chemical models; this inclusion introduces several free input parameters as the efficiencies of each process, especially the nonthermal desorption ones, are rather uncertain. Experiments on temperature desorption mechanisms now provide a clear indication of how species sublimate from the grains as a function of temperature.

Reaction databases: Ultimately the most important input in any model is the set of reactions to be adopted and their rate coefficients. The choice of reaction rate depends on factors such as availability, accuracy, and temperature ranges. As mentioned in Chapter 3, a number of comprehensive databases of rate coefficients are available today, with the most used being the UDfA, the Ohio State University, and the KIDA databases (see Further Reading for websites). They all collect the data from many sources, both theoretical and experimental.

9.1.2 What Comes out of the Chemical Models?

A typical chemical model will compute the fractional abundances of all atomic and molecular species included in the network, with respect to either the total number of hydrogen nuclei or molecular hydrogen. These fractional abundances will be given as a function of time (if the model is time-dependent) and space (if depth-dependent). If the model includes surface species then it will also provide these abundances for the ices. PDR models will also provide gas and dust temperatures, as well as the line emissivities and intensities of the main coolants, such as C, C^+, O, and CO. Ultimately, comparisons of grids of models where the input parameters have been varied with observations give information on physical parameters such as density, temperature, cosmic ray ionisation rates, and photorates.

9.1.3 PDR Modelling

Owing to the complexity of PDRs (see Section 3.2.1), it is worth describing models of PDRs in more detail. To characterise fully a photon-dominated region, all the local properties such as the fractional abundances of atoms and molecules, their level populations, the temperatures of the gas and dust, the gas pressure, and possibly the composition of dust and PAHs, etc., need to be computed. These computations are not trivial as photons penetrate the gas and hence different components of the clouds are effectively coupled. Where the radiation field is impinging and at what wavelength will affect the gas and dust heating, and hence the temperature balance and ultimately the chemistry.

The calculations of the temperature and the chemistry are closely inter-related, as the former will be, amongst other things, a function of the cooling by atoms and molecules. PDR codes typically iterate through the following computational steps: (1) they solve the local chemical balance to determine local densities; (2) they solve the local energy balance to estimate the local physical properties such as temperatures, pressures, and level populations; and (3) they solve the radiative transfer.

One of the key inputs in PDR models is the type of geometry. Most codes are one-dimensional and use a plane-parallel geometry, illuminated either from one side or from both sides. This geometry simplifies the radiative transfer problem significantly because it is sufficient to account for just one line of sight. Some codes use a spherical geometry.

The radiation field is one of the key parameters in PDRs, so we describe here how it is usually treated in PDR models. In most codes the ambient far UV is in units of the Draine or Habing radiation fields (defined in Section 9.1.1). For isotropic fields, the radiation is averaged over 4π (which assumes that the cloud is spherical). The main difference in a directed or an isotropic far UV is in the treatment of the dust attenuation. Dust scattering adds an extra complication to the attenuation of far UV radiation, because the albedo and phase function of the grains must be specified.

In terms of the chemistry, these are the main processes that are key to a proper PDR model and that require an accurate treatment: photoprocesses, the H/H_2 transition zone, and the $C^+/C/CO$ transition zone (see Figure 3.2).

To calculate the temperature of a cloud, equilibrium between heating and cooling has to be computed. The main sources of heating are (1) the H_2

formation (see Section 3.4.1) and destruction (see Section 3.2.1), (2) cosmic rays, (3) photoelectric heating, (4) X-ray heating, and (5) photoelectric effects from dust grains and PAH molecules, all of which ionise atoms and molecules and give energy to the free electron. The main sources of cooling are (1) atoms and molecules (O, C, C^+, CO, etc.), (2) gas–grain interactions, and (3) radiative recombination.

PDR models differ in the treatment of all these aspects, in particular, whether they use finite and semi-infinite plane-parallel, spherical geometry or disk geometry; treatment of the different heating and cooling mechanisms, steady-state chemistry versus time dependent, etc. However, benchmarking efforts are common among the PDR modellers, and references to available codes can be found in the Further Reading section at the end of the chapter.

9.1.4 Gas–Grain Modelling

Gas–grain models can be constructed using two conceptually different approaches: the first involves a simple or pseudo-surface chemistry based on the assumption that fast hydrogenation and/or oxidation reactions occur with high efficiencies until saturation, even if the temperatures are low (~ 10 K). For example, it might be assumed that O atoms are rapidly converted on the surface to H_2O, nitrogen to NH_3, and CO to H_2CO and CH_3OH. The advantage of this approach is that the surface reaction rate coefficients do not need to be specified: it is worth noting that such coefficients are very rarely available from experiments (see Section 3.4). The very few reaction coefficients that have indeed been investigated experimentally are usually extrapolated for interstellar conditions. They cannot easily be included in the same ODEs used for the gas phase because the dust provides a surface rather than a volume, and the former is not well known because dust grains may be highly porous. The disadvantage of this simple approach is, of course, that only a small set of surface reactions is included in these models (e.g., hydrogenation until saturation) and effectively 100% efficiencies in reaction are assumed.

The second approach is to include a complex surface reaction network in the model (see Section 3.4.3). Owing to the uncertainties in the rate coefficients for all the surface reactions as well as uncertainties concerning the nature of the surfaces of dust grains, several methods have been devised. There is the so called 'accretion-limited' method (through the use of rate equations), valid when the timescale for an atom or molecule to scan the surface is much less than the accretion time of the co-reactant. The chemistry is then limited by the accretion rate of new species. Another method is the 'reaction-limited' approach (through a Monte Carlo calculation) where the opposite holds true, so a species trapped in a site can react only with migrating species that visit that site. These methods, however, do not treat inhomogeneity of the dust grains. Recent developments to tackle surfaces being discontinuous are the special Monte Carlo 'continuous-time, random walk' approach and the modified rate equation approach.

9.2 Radiative Transfer Modelling

When local thermodynamic equilibrium (LTE) does not apply, it is necessary to consider all the individual processes that lead to level population or depopulation. Together, collisions and radiation determine the level populations through the equation of statistical equilibrium. The interpretation of molecular lines requires the use of line radiative transfer models able to calculate accurately the non-LTE level populations and the resulting output spectra. In particular, radiative transfer codes compute the flux or intensity of individual line emissions, and, in some cases the line profiles. The major prerequisite for a successful radiative transfer code is the availability and accuracy of collisional data.

Although the more general form for the radiative transfer equation and its simplified solution can be found in standard textbooks (see Further Reading), here we define an expression for spectral line emission in particular. Hence, we recall Equation 2.2:

$$\frac{dI_\nu}{d\tau_\nu} = -I_\nu + S_\nu \tag{9.2}$$

where I_ν is the radiation intensity as a function of frequency, and S_ν is the source function. The latter is equivalent to j_ν/α_ν, with j_ν being the emission coefficient and α_ν the absorption coefficient. It can be shown that:

$$j_\nu = \frac{h\nu}{4\pi} n_j A_{ij} \varphi(\nu) \tag{9.3}$$

and

$$\alpha_\nu = \frac{h\nu}{c}(n_j B_{ji} - n_i B_{ij})\varphi(\nu) \tag{9.4}$$

where A_{ij} and B_{ji} are, as before (see Chapter 2), the transition probability of spontaneous de-excitation and the transition probability of induced de-excitation, respectively; n_i and n_j are the number densities in levels i and j, respectively; and $\varphi(\nu)$ is the normalised profile function. The equation of statistical equilibrium for a transition $i \rightarrow j$ can be expressed as:

$$\frac{dn_i}{dt} = -n_i \left[\sum_{j<i} A_{ij} + \sum_{j\neq i}(B_{ij}J_\nu + C_{ij}) \right] + \sum_{j>i} n_j A_{ji} + \sum_{j\neq i} n_j (B_{ji}J_\nu + C_{ji})$$

(9.5)

where J_ν is the mean radiation field intensity, that is, the average I_ν over the solid angle, and C_{ij} the collisional rates. For a stationary situation, the molecular level populations do not change with respect to time and therefore:

$$n_i \left[\sum_{j<i} A_{ij} + \sum_{j\neq i}(B_{ij}J_\nu + C_{ij}) \right] = \sum_{j>i} n_j A_{ji} + \sum_{j\neq i} n_j (B_{ji}J_\nu + C_{ji}) \quad (9.6)$$

The collisional rates will depend on the density, temperature, and the collisional rate coefficients for each interaction between the molecule and its collisional partner. In non-LTE the Boltzmann equation (Equation 8.19) tells us that:

$$\frac{n_i}{n_j} = \frac{g_i}{g_j} \exp[-(h\nu)/(kT_{ex})] \quad (9.7)$$

Clearly, when collisions dominate then of course $T_{ex} \rightarrow T_k$ and we are in the LTE regime (see Section 8.2). So from Equations 9.6 and 9.7 the quantities that need to be known to solve the radiative transfer equation 9.2 are the excitation temperatures, which in non-LTE can be different for each level, and the radiation field that will have contributions from sources such as the Cosmic Microwave Background, dust emission, and spectral lines; the latter, of course, will depend on the relative populations of the different levels of a molecule.

9.2.1 Large Velocity Gradient Approximation

Often one can calculate the relative populations of different rotational states of the molecule from the equations of statistical equilibrium, with the radiative transfer being approximated by a simple expression for escape probability that involves the same populations. This approximation assumes the emitting region is homogeneous, isothermal, and without large-scale velocity fields, and allows us to decouple the radiative transfer and the statistical equilibrium equations.

Let us assume that a photon emitted at position r_1 can be absorbed at another position r_2 and will therefore influence the level population there; however, if we assume that the medium is homogeneous and isothermal then one can derive the probability of an absorption occurring while the photon is travelling from r_1 to r_2. This approximation is called the Sobolev approximation of large velocity gradient (LVG). In other words, LVG assumes that the velocity differences between r_1 and r_2 are large compared to the width of the velocity distribution at either r_1 and r_2. This, together with the assumption of uniform temperature, density, and abundance, ensures that the relative populations are the same throughout the cloud. An LVG calculation is different when it is dealing with linear or nonlinear molecules, but by including radiative and collisional transitions between arbitrary pairs of rotational levels one can handle both.

Mathematically, what this means is that we can estimate J_ν by defining a probability, β, that a photon emitted in a transition at position r_1 will escape from the cloud; then $J_\nu = S_\nu(1 - \beta)$. The column density is of course related to the optical depth of the medium as:

$$\tau_\nu = \int_0^D \alpha_\nu n_l(s)\,ds \qquad (9.8)$$

where τ_ν is the optical depth as a function of frequency, D is the distance over which absorption occurs, $n_l(s)$ is number density of the species in level l as a function of position s, and α_ν is the absorption coefficient. The integral of n_l over position gives the column density in level l.

The relation between $\beta (= 1/C_\tau)$ and τ in the Sobolev approximation is:

$$\beta = \frac{1 - e^{-\tau_\nu}}{\tau_\nu} \qquad (9.9)$$

for an expanding sphere, and:

$$\beta = \frac{1 - e^{-3\tau_\nu}}{3\tau_\nu} \qquad (9.10)$$

for a homogeneous slab. The mean radiation field can now be calculated independently of the level population. Equations 9.8, 9.9, or 9.10, and 9.7 form the basic equations of an LVG model.

When using an LVG code, there are three free parameters once the excitation rates are chosen: the kinetic temperature, T_k, of the H_2 molecules taking part in the collisional process; the number density of the hydrogen molecules, $n(H_2)$; and $n/(dv/dr)$, where n is the total number density of the molecule of interest and dv/dr the cloud velocity gradient.

The goodness of a fit with an LVG model can be assessed by a reduced χ^2 statistic:

$$\chi_f^2 = \frac{1}{f} \sum_{j=0}^{n} \left(\bar{T}_{Aj} - T_{Rj}^2/(\bar{\sigma}_j) \right)^2 \qquad (9.11)$$

where T_{Rj} is the brightness temperature predicted by the LVG model for transition j, $\bar{T}_{Aj}$ is the weighted average of the observed antenna temperatures for this transition, and $\bar{\sigma}_j$ is the uncertainty in this average; the sum is over all the observed transitions and f is the number of degrees of freedom.

The collisional coefficients represent the major sources of uncertainty for many molecules, of course, but even assuming that these are known we must not forget that LVG models have an intrinsic degeneracy in that often the combination of gas density and temperature that gives the best LVG fit does not necessarily imply the right density or temperature. Moreover, once observations tell us that there is a clear structure present in the region, the LVG approximation is, by definition, deficient.

Examples of available LVG codes are given in the Further Reading section.

9.2.2 Non-LVG Approach

The LVG approach is clearly an approximation and, although very powerful in constraining the parameter space of possible densities and temperatures that fit the observational data, at times one may need to solve the radiative transfer problem more accurately.

Several techniques can be employed to treat the radiative transfer problem, and it is well known, especially from stellar atmosphere modelling, that approximations in the radiative transfer algorithms can result in substantial errors in the line intensity estimates.

As we have seen in the previous section, the radiative transfer problem can be summarised by an equation describing the emission, absorption, and movement of photons along a straight line in a medium. If LVG is not assumed then the set of equations that calculate the radiation field and the level populations need to be solved simultaneously, usually via an iterative process. A sketch of the iteration process is shown in Figure 9.1 (recall Section 2.2).

Different methodologies can be employed to do this. Among the most common ones we have:

1. Lambda Iteration (LI): This relies on the iterative evaluation of level populations and intensities until the system has converged.
2. Monte Carlo (MC): This method simulates the basic physical processes with the aid of random numbers.
3. Approximated/Accelerated LI (ALI): As LI but the equations are preconditioned to speed up convergence.
4. Accelerated MC (AMC): As MC but some pre-conditioning occurs.

The input parameters required, apart from the collisional data, are (1) the species abundances, (2) density and temperature as a function of depth into the cloud, (3) radiation field, and (4) the velocity field. Among the most common

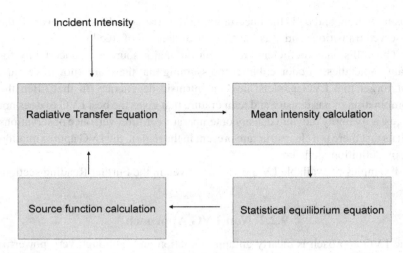

Figure 9.1. Flow diagram of the molecular line radiative transfer problem. (Re-drawn from van Zadelhoff, G.-J., Dullemond, C. P., van der Tak, F. F. S., Yates, J. A., Doty, S. D., Ossenkopf, V., Hogerheijde, M. R., Juvela, M., Wiese-meyer, H., and Schier, F. L. 2002. *Astronomy & Astrophysics*, 395, 373.)

outputs are the level populations, the excitation temperatures, the line intensi-ties, and line profiles (although not all codes deliver the latter).

There have been several works in the recent literature that couple chemical models (providing (1) and (2)) with radiative transfer models. As an example, here we summarise the findings of a theoretical study of the $^{12}C^{32}S$ (hereafter CS) molecular emission that aimed at providing observers with theoretical CS profiles to interpret the emission from this molecule in hot cores. In this work a chemical model was run for a large parameter space covering wide ranges of core sizes (which included an ultracompact core and an envelope), densities, temperatures, and ages. The outputs of this grid were then fed into a radiative transfer code based on ALI, which considered the first 40 rotational levels of CS in the vibrational ground state; the molecular data and the collisional rates with respect to H_2 are from the Cologne Database for Molecular Spectroscopy (see Further Reading). The quantities calculated by the radiative transfer code were the intensities and emission spectra. The latter were also then convolved with the appropriate telescope beam (in this case either JCMT or IRAM) and estimates of line fluxes at various resolution were finally provided. Figure 9.2 shows an example of this output.

Beside the value of this type of work for future ALMA observations, one can also draw some general astrochemical conclusions; for example, it was found that the CS fractional abundance is highest in the innermost parts of

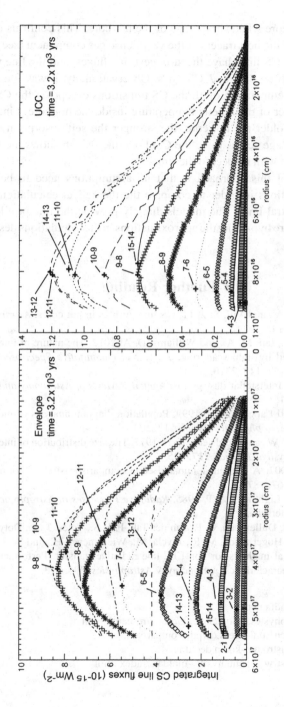

Figure 9.2. Variation of the CS integrated line fluxes for an ultracompact core model (right-hand side) and its envelope (left-hand side). (Reproduced by permission of the AAS from Bayet, E, Yates, J. A., and Viti, S. 2011. *Astrophysical Journal*, 728, 114.)

the ultracompact core whatever the age of the hot core. This confirms the CS molecule as one of the best tracers of the very dense gas component (see Chapter 5). The high-J CS lines have the strongest line fluxes, and the line widths are broader than those of low-J CS lines. Observationally, it was found that the peak antenna temperature of all the CS transitions except for the CS(1–0) line is a good tracer of the kinetic temperature inside the hot core. Finally, in the envelope, the older the hot core, the stronger the self-absorption of CS. The best tracer of age seems to be the CS(1–0) line, which shows the largest variations in fluxes with respect to the time.

The neatness of this approach is that no assumptions need to be made about the observations in order to compare the theoretical calculations with the observed spectral lines. The models directly provide the line profiles and intensities, therefore bypassing the approximations in the calculations described in Chapter 8.

9.3 Further Reading

Bayet, E., Yates, J. A., and Viti, S. 2011. CS line profiles in hot cores. *Astrophysical Journal*, 728, 114.

Cecchi-Pestellini, C., Iati, M. A., and Williams, D. A. 2012. The nature of interstellar dust as revealed by light scattering. *Journal of Quantitative Spectroscopy and Radiative Transfer*, 113, 2310.

Draine, B. T. 2003. Interstellar dust grains. *Annual Reviews of Astronomy and Astrophysics*, 41, 241.

Goldsmith, P. F., and Langer, W. D. 1999. Population diagram analysis of molecular line emission. *Astrophysical Journal*, 517, 209.

Mathis, J. S., Rumpl, W., and Nordsieck, K. H. 1977. The size distribution of interstellar grains. *Astrophysical Journal*, 217, 425.

Röllig, M. et al. 2007. A photon dominated code comparison study. *Astronomy & Astrophysics*, 467, 187.

Rybicki, G. B., and Lightman, A. P. 1985. *Radiative Processes in Astrophysics*. New York: Wiley-Interscience.

van Zadelhoff, G.-J., Dullemond, C. P., van der Tak, F. F. S., Yates, J. A., Doty, S. D., Ossenkopf, V., Hogerheijde, M. R., Juvela, M., Wiesemeyer, H., and Schier, F. L. 2002. Numerical methods for non-LTE line radiative transfer: Performance and convergence characteristics. *Astronomy & Astrophysics*, 395, 373.

Websites:

- http://www.udfa.net/
- http://www.physics.ohio-state.edu/~eric/research.html
- http://www.ph1.uni-koeln.de/pdr-comparison.
- http://www.astro.uni-koeln.de/cdms/
- http://home.strw.leidenuniv.nl/~moldata/radex.html

10

Observations: Which Molecule, Which Transition?

The preceding chapters in this book have demonstrated that to trace particular astronomical features in the Milky Way or in external galaxies by using molecular line emissions, the astronomer needs to choose lines corresponding to appropriate transitions. The transitions to use will, obviously, be those whose upper levels are readily populated in the gas that is to be observed. In many situations, the most important excitation mechanism to the upper level is collisional, and H_2 is often the main collisional partner.

For example, we have seen that the CO(1–0) transition is appropriate for searching for and detecting cold neutral gas with a kinetic temperature of $\sim$10 K, where the number density of hydrogen molecules is $\sim 10^3$ cm^{-3}. However, observations of radiation emitted in this transition cannot reveal, say, the presence of either cold or warm gas at a density of, say, $\sim 10^5$ cm^{-3}, because collisional de-excitation of the upper level occurs before radiation in the (1–0) line can occur. Therefore, to observe gas at higher densities, observers must use more highly excited CO lines that have larger spontaneous radiation probabilities (assuming that these highly excited levels are sufficiently populated at the prevailing temperature). Alternatively, observers may use a line from some other molecular species that has more appropriate fundamental properties for the physical conditions in the gas to be observed. Of course, as we have seen in Chapters 8 and 9, complications introduced by high optical depths in the lines observed may also make it difficult to infer physical properties in the observed regions. The simple physics in the above arguments is encapsulated in the concept of critical density (see Section 2.3).

There are, currently, more than 180 different gas-phase molecular species so far detected in interstellar and circumstellar regions of the Milky Way, and a growing number of molecular species (currently more than 50) have also been detected in external galaxies. Some of these molecular species have been detected in many transitions. In an extreme example, CO has been detected in

165

Table 10.1. Examples of critical densities for selected species at specific temperatures

Molecule	Formula	Transition	ν (GHz)	E_u (K)	n_{crit} (cm^{-3})	T(K)
Carbon monoxide	CO	(1–0)	115.27	5.53	1.8×10^3	10
Carbon monoxide	CO	(6–5)	691.47	116.16	2.5×10^5	100
Carbon monosulfide	CS	(4–3)	195.95	23.51	2.6×10^6	20
Carbon monosulfide	CS	(9–8)	440.80	105.79	4.3×10^7	100
Carbonyl sulfide	OCS	(8–7)	97.30	21.01	3.5×10^4	20
Carbonyl sulfide	OCS	(18–17)	218.90	99.81	4.0×10^5	100
Sulfur monoxide	SO	(5_6-4_5)	219.95	34.98	3.5×10^6	50
Sulfur monoxide	SO	(8–8)	254.57	99.7	2.8×10^6	100
Sulfur dioxide	SO$_2$	$(2_{20}-2_{11})$	151.38	8.75	8.7×10^6	10
Sulfur dioxide	SO$_2$	$(7_{44}-7_{35})$	357.39	69.47	4.4×10^7	75
Silicon monoxide	SiO	(3–2)	130.27	12.50	8.1×10^5	10
Silicon monoxide	SiO	(9–8)	390.73	93.77	2.9×10^7	100
Formyl cation	HCO$^+$	(2–1)	178.38	12.84	1.1×10^6	10
Formyl cation	HCO$^+$	(7–6)	624.21	119.84	4.9×10^7	100
Diazenylium	N$_2$H$^+$	(2–1)	186.34	13.41	9.2×10^5	10
Diazenylium	N$_2$H$^+$	(7–6)	652.09	125.19	4.1×10^7	100
Cyanoacetylene	HC$_3$N	(12–11)	109.17	34.06	7.1×10^5	20
Cyanoacetylene	HC$_3$N	(19–18)	172.85	82.96	2.9×10^6	80
Hydrogen cyanide	HCN	(2–1)	177.26	12.76	1.0×10^7	10
Hydrogen cyanide	HCN	(7–6)	620.30	119.09	1.2×10^9	100
Hydrogen isocyanide	HNC	(3–2)	271.98	26.11	8.0×10^6	10
Hydrogen isocyanide	HNC	(7–6)	543.89	91.37	7.6×10^7	100
Formaldehyde	p-H$_2$CO	$(2_{02}-1_{10})$	145.60	10.48	1.1×10^6	10
Formaldehyde	p-H$_2$CO	$(3_{22}-2_{21})$	218.47	68.00	6.8×10^6	70
Formaldehyde	o-H$_2$CO	$(2_{12}-1_{11})$	140.84	21.9	8.8×10^5	20
Formaldehyde	o-H$_2$CO	$(5_{15}-4_{14})$	351.77	62.445	1.8×10^7	70
Methanol	e-CH$_3$OH	$(2_{00}-1_{00})$	96.75	12.19	2.8×10^5	10
Methanol	e-CH$_3$OH	$(7_{30}-6_{30})$	338.58	104.81	2.0×10^7	100
Methanol	a-CH$_3$OH	$(3_{00}-2_{00})$	145.10	13.93	1.8×10^5	10
Methanol	a-CH$_3$OH	$(7_{-20}-6_{-10})$	812.55	102.70	1.9×10^8	100
Deuterated water	HDO	$(1_{01}-0_{00})$	464.92	22.3	3.0×10^6	20
Deuterated water	HDO	$(2_{11}-2_{12})$	241.56	95.2	1.1×10^5	100
Isocyanic acid	HNCO	$(5_{05}-4_{04})$	109.90	15.8	1.0×10^7	20
Isocyanic acid	HNCO	$(15_{015}-14_{014})$	329.66	126.6	1.2×10^8	80
Nitric oxide	NO	$(3_{13}-2_{-12})$	250.44	19.23	2.4×10^5	10
Nitric oxide	NO	$(7_{18}-6_{-17})$	752.01	151.63	4.8×10^7	150

Where available the collisional partner used to calculate the critical densities is H$_2$ *ortho*. The table lists the molecule and chemical formula, the selected transitions, the frequency in GHz, the energy above ground of the upper level, and the critical density, n_{crit} (cm^{-3}), computed at a selected temperature T(K).

pure rotational transitions from all the levels from $J = 1$ up to $J = 37$ in the ground vibrational level. However, most molecules have been detected in just one or a few transitions, but this still leaves plenty of choice when planning an observational programme. It is important to try and select particular molecular lines from the thousands of possible choices that will reveal most easily the properties of the interstellar or circumstellar regions that are to be studied.

Not all of the possible molecular lines have equal value to the astronomer as tracers of the interstellar or circumstellar gas. In Table 10.1 we summarise some useful information about a selection of transitions of gas-phase molecular tracers that have been and continue to be influential in molecular astrophysics. The selection is arbitrary but illustrates how appropriate molecular lines can probe regions of widely differing physical conditions.

Table 10.1 is based on data from the Leiden Atomic and Molecular Database (LAMDA) (see Further Reading). The database refers to collisions in which the collision partner is either H_2(*para* and *ortho*) or He. We have arbitrarily selected transitions in the range 100–1000 GHz (appropriate for ALMA) and compute critical densities at two temperatures, usually 10 K and 100 K, for each of these transitions. We restrict to transitions in the ground vibrational level, and – apart from HDO – we ignore isotopic substitutions.

Table 10.1 lists the following information for each molecular transition, and each species has two lines in the table: (1) name of molecule, (2) chemical formula, (3) transition, (4) frequency, (5) energy of upper level (K), and (6) critical density at a specific temperature.

10.1 Further Reading

Schoier, F. L., van der Tak, F. F. S., van Dishoeck, E. F., and Black, J. H. 2005. An atomic and molecular database for analysis of submillimetre line observations. *Astronomy & Astrophysics*, 432, 369.
An updated version of this paper is maintained at the website home.strw.leidenuniv.nl/~moldata/

Appendix

Acronyms

AGB	Asymptotic Giant Branch
AGN	Active Galactic Nucleus
ALI	Accelerated Lamda Iteration (method)
ALMA	Atacama Large Millimeter Array
AMC	Accelerated Monte Carlo (method)
AURA	Association of Universities for Research in Astronomy
BIMA	Berkeley Illinois Maryland Association
CDM	Cold Dark Matter
CMB	Cosmic Microwave Background
CMZ	Central Molecular Zone (of the Milky Way Galaxy)
COM	Complex Organic Molecule
CRDR	Cosmic Ray–Dominated Region
CSE	Circumstellar Envelope
ESA	European Space Agency
FCRAO	Five College Radio Astronomy Observatory
FUSE	Far Ultraviolet Spectroscopic Explorer
GMC	Giant Molecular Cloud
HHO	Herbig–Haro Object
HIFI	Heterodyne Instrument for the Far Infrared (of the Herschel Space Observatory)
IMF	Initial Mass Function
IRAS	InfraRed Astronomical Satellite
IRDC	InfraRed Dark Cloud
ISEC	InterStellar Extinction Curve
JWST	James Webb Space Telescope
KIDA	Kinetic Database for Astrochemistry
KL	Kleinmann–Low (complex)
LAMDA	Leiden Atomic and Molecular Database
LI	Lamda Iteration (method)
LMH	Large Molecule Heimat
LTE	Local Thermodynamic Equilibrium
LVG	Large Velocity Gradient (approximation)
MC	Monte Carlo (method)

MRN	Mathis–Rumpl–Nordsieck (grain size distribution)
MWG	Milky Way Galaxy
NASA	National Aeronautics and Space Administration
ODE	Ordinary Differential Equation
PAH	Polycyclic Aromatic Hydrocarbon (molecule)
PdBI	Plateau de Bure Interferometer
PDR	Photon-Dominated Region
PN	Planetary Nebula
PPN	Proto-Planetary Nebula
SKA	Square Kilometer Array
SMA	Submillimeter Array
SN	Supernova
STScI	Space Telescope Science Institute
UCHII	Ultracompact HII (region)
UDFA	UMIST Database For Astrochemistry
ULIRG	Ultra-Luminous InfraRed Galaxy
XDR	X-ray–Dominated Region

Index